Research Reports in Physics

Research Reports in Physics

Nuclear Structure of the Zirconium Region
Editors: J. Eberth, R. A. Meyer, and K. Sistemich

Ecodynamics Contributions to Theoretical Ecology
Editors: W. Wolff, C.-J. Soeder, and F. R. Drepper

Nonlinear Waves 1 Dynamics and Evolution
Editors: A. V. Gaponov-Grekhov, M. I. Rabinovich, and J. Engelbrecht

Nonlinear Waves 2 Dynamics and Evolution
Editors: A. V. Gaponov-Grekhov, M. I. Rabinovich, and J. Engelbrecht

Nonlinear Waves 3 Physics and Astrophysics
Editors: A. V. Gaponov-Grekhov, M. I. Rabinovich, and J. Engelbrecht

Nuclear Astrophysics
Editors: M. Lozano, M. I. Gallardo, and J. M. Arias

Optimized LCAO Method and the Electronic Structure of Extended Systems
By H. Eschrig

Nonlinear Waves in Active Media
Editor: J. Engelbrecht

Problems of Modern Quantum Field Theory
Editors: A. A. Belavin, A. U. Klimyk, and A. B. Zamolodchikov

Fluctuational Superconductivity of Magnetic Systems
By M. A. Savchenko and A. V. Stefanovich

Nonlinear Evolution Equations and Dynamical Systems
Editors: S. Carillo and O. Ragnisco

Nonlinear Physics
Editors: Gu Chaohao, Li Yishen, and Tu Guizhang

Nonlinear Waves in Waveguides with Stratification
By S. B. Leble

Quark-Gluon Plasma
Editors: B. Sinha, S. Pal, and S. Raha

Symmetries and Singularity Structures Integrability and Chaos
in Nonlinear Dynamical Systems
Editors: M. Lakshmanan and M. Daniel

Modeling Air-Lake Interaction Physical Background
Editor: S. S. Zilitinkevich

Sergei B. Leble

Nonlinear Waves in Waveguides

with Stratification

With 6 Figures

Springer-Verlag
Berlin Heidelberg New York London Paris
Tokyo Hong Kong Barcelona Budapest

Professor Dr. Sergei B. Leble
Department of Theoretical Physics, Kaliningrad State University,
Aleksandr Nevskii St. 14, SU-236041 Kaliningrad, USSR

ISBN 3-540-52149-6 Springer-Verlag Berlin Heidelberg New York
ISBN 0-387-52149-6 Springer-Verlag New York Berlin Heidelberg

This work is subject to copyright. All rights are reserved, whether the whole or part of the material is concerned, specifically the rights of translation, reprinting, reuse of illustrations, recitation, broadcasting, reproduction on microfilms or in other ways, and storage in data banks. Duplikation of this publication or parts thereof is only permitted under the provisions of the German Copyright Law of September 9, 1965, in its current version, and a copyright fee must always be paid. Violations fall under the prosecution act of the German Copyright Law.

© Springer-Verlag Berlin Heidelberg 1991

Printed in Germany

The use of registered names, trademarks, etc. in this publication does not imply, even in the absence of a specific statement, that such names are exempt from the relevant protectiv laws and regulations and therefore free for general use.

57 / 3140-543210 – Printed on acid-free paper

Preface

This book emerged from a special course on "Nonlinear Wave Theory" at the Kaliningrad State University. An exhaustive overview of research into nonlinear wave motion in waveguides, in particular, in the USSR, is given. The theoretical methodology in describing finite amplitude waves is emphasized and illustrated by the author's own results. The basic element of this method is the formalization of the derivation of the nonlinear evolution system (NES), i.e., the simplification of the fundamental physical equations. The procedure entails the expansion in powers of small parameters that characterize the initial or boundary conditions. In other words, the sets of initial (boundary) conditions are described in such a way that the initial physical problem may be transposed to an NES.

The main concepts of the method have been formulated as the generalization of the theory of nonlinear oscillations derived by Poincaré and Bogolyubov. The estimation of the magnitudes of the appearing expansion terms and the time interval in which the asymptotic (in the small parameter) representation of the evolution is valid include information on the NES solution as well as an assumption about the initial values of the dynamical variables and the amplitudes of their derivatives. If the interacting waves can be separated according to their type, then the NES becomes extremely simple in terms of the roots of the linearized dispersion equation. This wave separation requires the introduction of a special technique of projection onto the subspaces of the solutions of the dispersion branches.

We have decided to restrict ourselves to the waveguide propagation of nonlinear waves. This problem is a convenient example of a non-one-dimensional boundary problem. A discrete transverse component in the wave vector allows the introduction of the concept of interacting modes. The analogy to the Bubnov–Galerkin method and the approximate description of finite number level in quantum mechanics can also be shown. The projection procedure with mode expansion allows to represent the nonlinear evolution in a waveguide as the result of the interaction of elementary modes.

Propagation in a nonlinear waveguide has very broad applications, for example, fiber optics, where generation and transmission of extremely short pulses is important, toroidal plasma physics and ionospheric waveguides, oceanic and atmospheric acoustic-gravitational waves. These various (physical) phenomena have many common mathematical features.

Collaboration with geophysicists has led to important applications of these new methods to internal gravity waves in the atmosphere and ocean for which significant space is reserved in this book. We note that in the upper atmosphere (in the thermosphere) specific waveguide propagation conditions occur. The collisionless regime and the action of the gravitational field restrict the atmospheric waves to a limited height interval.

The theoretical basis of propagation in waveguides is quite naturally demonstrated in the case of electromagnetic waves, especially in such classical problems as metal tube, coaxial cable and dielectric rod (fiber) waveguides. The existing high resolution technique of finding the parameters of electromagnetic waves may help to investigate the effects that are related to complicated single-mode propagation and multi-mode interaction phenomena. The theory of turbulent plasma is studied in connection with Silin-Tikhonchuk solitons. The derivation of NES for plasma waves requires a generalization of the method, i.e., the expansion in small parameters of an ion distribution function.

The variety of types of plasma waves that can be considered represents the broad applications of the method of elementary modes. The special problems for the NES of any type of wave are accompanied by widely applicable solutions that allow to approximate a given boundary regime. For the solution of the near-integrable systems, nonsingular perturbation theory is applied just as in the NES derivation.

The author is indebted to the Leningrad State University for providing the possibility to contemplate and discuss the principal features of these investigations. The criticism of the theoretical physics department, especially of Professors Buldyrev and Buslaev, helped to clarify many details. Professor Buldyrev read part of the text of this book and made many useful remarks. I am grateful to these two gentlemen as well as to S.Yu. Dobrokhotov, A.A. Zaitsev, V.B. Matveev, A.A. Namgaladze and the colleagues from the theoretical physics departments of Leningrad and Kaliningrad Universities for productive discussions. For their help with the manuscript I would like to thank N. Aristov and J. Engelbrecht and in particular E. Klement who transformed my manuscript into such a nice form. In favor of the most important feature of "timeliness", the English had to be left in a form still containing a bit of Russian flavor.

Kaliningrad November 1990 *S. Leble*

Contents

1. **Introduction** 1
   1.1 General Remarks 1
   1.2 Overview 8

2. **The Discrimination and Interaction
   of Hydrodynamical Waves of Different Types** 12
   2.1 Hydrodynamic Wave Equations in a Three-Dimensional
       Stratified Medium 12
       2.1.1 Fundamentals 12
       2.1.2 Atmospheric Waves: Dispersion and Polarization 15
   2.2 Projection Operators on Subspaces of Waves 19
       2.2.1 Determination of Wave Type 19
       2.2.2 Projection Operators in Geophysical Hydrodynamics .. 19
       2.2.3 Projection Operators for Poincaré and Rossby Waves . 21
   2.3 Interaction of Surface Poincaré and Rossby Waves 24
   2.4 Long Atmospheric Internal Waves 26
       2.4.1 Internal Waves in an Atmospheric Waveguide 27
   2.5 The Method of Successive Approximations.
       Solution for the Cauchy Problem
       of the Coupled Korteweg-de Vries Equations 37
   2.6 The Coupled KdV Equations 40
       2.6.1 Quasisolitons of CKdV 40
       2.6.2 The Symmetry and Integrability of CKdV 41
       2.6.3 The Stationary Solutions of CKdV 44
   2.7 Nonlinear Rossby Waves on a Sphere 45

3. **Interaction of Modes in an Electromagnetic Field Waveguide** ... 50
   3.1 Electromagnetic Fields in Dielectric Waveguides 50
       3.1.1 Basic Equations and Boundary Conditions 50
       3.1.2 Waveguide Modes 52
   3.2 Equations of Mode Interaction 55
   3.3 Three-Wave Interaction of Magnetic Modes
       in an Extended System of Nuclear Spins 58
   3.4 Nonlinear Mode Dispersion in Dielectric Waveguides 64

4. Nonlinear Waves in Stratified Plasma ... 69

- 4.1 Wave Modes in Stratified Plasma ... 69
  - 4.1.1 Description of Plasma Waves ... 69
  - 4.1.2 Weakly Inhomogeneous Bounded Plasma ... 74
- 4.2 Interaction of Plasma Waves ... 76
  - 4.2.1 Langmuir and Ion Wave Interactions ... 76
  - 4.2.2 Three-Wave Helico-Acoustic Interaction ... 79
- 4.3 Statistical Averaging and Weak Turbulence in Plasma ... 82
  - 4.3.1 Spectral Densities of Electromagnetic Field Fluctuations ... 82
  - 4.3.2 One-Dimensional Turbulence ... 85
- 4.4 Multisoliton Solutions of Discrete Silin–Tikhonchuk Equations ... 88
  - 4.4.1 The Darboux–Matveev Transformation ... 88

5. Evolution Equations for Internal Waves in Media with Strongly Inhomogeneous Stratification ... 93

- 5.1 Waves in a Medium with Varying Stratification ... 93
- 5.2 Analogues of the KdV and KP Equations with Nonlocal Dispersion ... 98
- 5.3 Quasi-Waveguide Propagation of Internal Waves in the Atmosphere ... 104
- 5.4 Waves in Gases Inhomogeneous in Knudsen Number ... 106

6. Mean Field Generation by Waves in a Dissipative Medium ... 114

- 6.1 Description of a Wave-Medium System by the Scales of Motion ... 114
- 6.2 Thermoactivity and Diffusion in the Presence of Waves ... 119
- 6.3 Reaction of Ionosphere Plasma to a Passing Internal Wave ... 123
- 6.4 Interaction of Internal Waves with a Dissipative Medium ... 127
  - 6.4.1 Temperature Field Instability Induced by an Internal Wave ... 127
  - 6.4.2 Nonsingular Perturbation Theory ... 132
- 6.5 Mean Field Generation in a Dissipative Medium and Fine Structure of the Oceanic Thermoclyne ... 135

Appendix 1 Atmospheric Waves over a Rotating Planet ... 139

Appendix 2 Nonlinear Terms for Interacting Modes of Poincaré and Rossby Waves at a Rotating Channel Surface ... 141

Appendix 3 Projection to the Eigen Subspaces for Acoustic and Internal Waves ... 143

Appendix 4 Basis Vectors, Interaction Operator for Atomic Nuclei
with Spin 1 145

Appendix 5 Nonlinear Internal Waves
and Shear Flow Interaction Constants 148

References 151

Subject Index 161

1. Introduction

We shall first give some general remarks related to history, physical applications and general context. An overview of the material presented in this book follows.

1.1 General Remarks

In this book the apparatus of the model nonlinear evolution equations and the theory of wave interactions in waveguides (quasi-waveguides) that result from an inhomogeneity in the propagation medium are presented. The theoretical description of finite amplitude wave dynamics naturally concerns problems in mathematical physics as well as geophysical hydrodynamics [1.1–3]: The development of the theory of wave motion was motivated by the need for a description of surface waves. Surface wave propagation, however, may be considered as a limiting degenerate case of waveguide propagation [1.4]. Within this framework, it is possible to describe the system by choosing one or more coordinates in such a manner that the dependence of dynamical variables along them may have wave structure but no propagation. These coordinates are referred to as transverse; the remaining ones we shall call longitudinal.

The origins of the theory of nonlinear waves lie in the works of the previous century by *J.Scott–Russell, D.Korteweg, G.de Vries, J.Boussinesq, J.Stokes* [1.5–7]. In these papers the fundamental ideas of the solitary waves (*Scott–Russell*), counteracting nonlinearity and dispersion effects (*Stokes*), and the distinction between waves moving in opposite directions (*Boussinesq*) were introduced. In addition, the approximate (nonlinear) dispersion relations were derived here.

The "new wave" of interest in finite amplitude wave theory was initiated by the development of the inverse problem method in 1967 and the rapid progress of the solutions of the integration of the model evolution equations. The success of the theory includes the development of experimental soliton physics and the investigation of organized nonlinear phenomena. Examples are investigations of surface and internal waves in hydrochannels and rotating vessels [1.8,9], and modelling of nonlinear wave phenomena in transmission lines with a given nonlinearity, dispersion and dissipation [1.10]. New discoveries in electromagnetic nonlinear wave physics [1.11] led to new technological devices such as the splendid example of one-mode solitons propagating in fiber optic waveguides [1.12]. Progress in the design of powerful sources and control devices that are necessary for investigations of large-amplitude media disturbances allowed the global control of geophysical parameters. These experiments provided ample information for theoretical interpretation.

As a result, a gap between the development of the physics and the mathematical methods has formed. In one-dimensional problems, noticeable progress has been made through the efforts of many authors [1.12–15], but the essentially multi-dimensional nonlinear wave theory is not at all complete. The propagation of waves in guides is one such problem. We should add that the rapid perfection of numerical methods and computers allows us to solve complex problems that model the physical situation more and more exactly. Development of analytical techniques combined with numerical simulation allow mutual testing and broaden the regions of validity of the theory.

The elaborate methods of reduction of equation systems introduce new model evolution equations into mathematical physics. Thus, the classic waveguide propagation theory for surface and internal waves has given birth to a variety of integrable nonlinear equations – Korteweg–de Vries (KdV), Kadomtsev–Petviashvili (KP), Nonlinear Schrödinger (NS), Benjamin–Ono (BO), Joseph, Johnson, etc., the universal applicability of which has influenced all areas of physics.

This book may be divided into two parts. The first is devoted to the derivation of the model evolution equation for electromagnetic, atmospheric, water and plasma waves as needed for the solution of problems of wave propagation and interaction (Chaps. 2–5). In the second, wave effects such as nonlinear generation of non-wavelike disturbances are considered in the presence of dissipation and diffusion processes (Chap. 6).

In the first part we emphasize the investigation of the system of coupled KdV equations. The system demonstrates the basic tool of the new description of two- and three-dimensional nonlinear waves in guides. By means of the independent variables in the first order linear approximation the basic wave mode is introduced. This mode is a projection onto the eigen subspace of a Sturm–Liouville problem for transverse variables. The desired quantity w is represented by a Fourier series

$$w = \sum Z_n(z)\theta_n(x,t) \tag{1.1}$$

where z is the transverse and x the longitudinal coordinate. The sum in (1.1) is due to the discreteness of the Sturm–Liouville spectrum. It is this property that leads to waveguide propagation. In the particular case of long waves the mode coefficient functions $\theta_n(x,t)$ satisfy the KdV coupled equations (CKdV)

$$\theta_{nt} + c_n\theta_{nx} + \sigma \sum_{m,k} \Phi_n^{mk} \theta_m \theta_{kx} + \beta^2 M_n \theta_{nxxx} = 0, \quad n = 0, 1, 2, \ldots \tag{1.2}$$

where Φ_n^{mk}, M_n are nonlinear and dispersion constants. The parameters σ, β characterize the amplitude and scale of an initial condition.

Let us turn to the role of the boundary conditions that determine the basic functions Z_n for the expansion (1.1). In the simple cases of classic waveguides, these are the uniform conditions that give the standard spectral problem for a finite interval. If otherwise, due to significant inhomogeneity in the stratification in z a quasi-waveguide is formed and the waves are captured in any transverse coordinate interval, the boundaries are conventional. This leads to a change in the dispersion type, usually it is nonlocal. The problems that lead to model equations with such pseudodifferential operators are considered in Sect. 2.4 and Chap. 4. The study of interaction of waves with a nonlocal dispersion law is one of the current directions in waveguide propagation theory. It includes the derivation and investigation of BO and Joseph equations as well as their short wave, two-dimensional and multi-mode generalization. Separating the propagation medium into regions of varying stratification scales improves the convergence of the expansions and allows one to get a compact representation for the quasi-waveguide propagation solutions.

The application of Fourier analysis and perturbation theory to nonlinear oscillations began with the well-known works of A.Poincaré, N.N.Bogolyubov and B.G.Galerkin. The generalization of their results was done by *Taniuti* [1.16–19], *Maslov, Dobrokhotov* [1.20–22], *Grimshaw* [1.23–24], *Ostrovsky* [1.24–25] and *Pelinovsky* [1.4,25]. The development of the concept of interacting guide modes was continued by *Miropolsky* [1.26], *Pelinovsky* [1.25] and this author [1.27–28]. The difficulty of the problem lies in the ambiguity of the correct choice of the sequence in the perturbation technique for solving the nonlinear equation for which the simple perturbation theory shows a singularity (instability, effect accumulation). The presence of one or two space variables obviously complicates the description. Moreover, the question arises whether the expansion (1.1) converges at all orders of amplitude and dispersion parameters. The answer could be obtained only from the arbitrary mode decomposition term. Therefore it is necessary to explore the dependence of the solution on the same small parameters used to derive this sequence of evolution systems. This is the program of our book.

We now give a short review of the results that are directly related to this approach. The appropriate CKdV system has been introduced in [1.29] and discussed in [1.30]. Another version of this system as well as single- and two-soliton solutions have been given by *Hirota* and *Satsuma* [1.31] without any discussion of the physical implications. The general CKdV have been derived here for different hydrodynamical systems (Sects. 2.1–4) [1.27,32]. The explicit form for nonlinear and dispersion constants was obtained. In a collaboration with Kshevetsky the approach to CKdV integration was developed. This allows one to reduce the wave disturbance solution, which is nearly a single-mode wave, to the solution of a combined KdV-MKdV equation [1.33] (by the same amplitude parameter as in the derivation of CKdV) which is integrable [1.1] (Sect. 2.5,6). The solution of the latter equation and formulas for intermode couplings give the representation of the multi-mode problem. The amplitudes of mode contributions decrease rapidly enough with mode number n, so that convergence of the expansion (1.1) is obtained. In [1.32,34] the limiting case $M_n \to 0$ of the CKdV (1.2) was considered. The solutions have been constructed with characteristics given in the same approximation as that of the perturbation scheme. The analysis of the solution demonstrates the tendency to a self-localization due to wave mode interaction (Sects. 2.5,6). Independent investigations were carried out which showed that in the interaction of two Riemann waves similar results are obtained [1.35].

The integrability of the CKdV system was studied using the Lax (Sect. 2.6) and Wahlquist–Estabrook methods [*Kshevetsky* and *Leble,* unpublished]. It is shown that there exist two-mode systems generalizing the Hirota–Satsuma equation that have a Lax pair. Independently, *Dodd* and *Fordy* found an L-A pair for the Hirota–Satsuma equations [1.36].

The methods of nonlinear evolution equations, including waveguide systems, are presented in [1.4]. The material in this volume is similar to the first stage of our method which also follows *Ostrovsky* and *Pelinovsky* [1.24,25]. The next stages, the interactions in small terms and dispersion branches separation, have not been pursued there, however. The simplest evolution equation has been derived in [1.4] for a single mode case. The development of nonlinear evolution equations and progress in the theory of integrable systems [1.37] as well as the discovery of integrable systems (e.g., the Benney–Kaup and Ito systems [1.38]) allows one to hope for further propagation of the method.

The important stage of the simplification of the description of nonlinear wave dynamics is the solution of the dispersion branch determination (i.e, separation of waves according to their types). The formulation of this problem appeared in the pioneering work [1.6] in a search for a connection between the solutions of Boussinesq and KdV equations. The traditional derivation of the KdV equation is done by separating the "left" and "right" waves through the introduction of a small parameter σ in the wave function argument, e.g., $\theta(\sigma t, x - ct)$. This allows one to get the equation for a single-directed wave but does not give a unique form for interaction terms. Traditionally, separation methods for an individual case follow the characteristic (often spectral) properties of a given branch. The conditions prohibiting another branch might be transversality in the case of an electromagnetic wave or incompressibility in the case of an internal gravity wave in a stratified fluid. The wave separation decreases the order of the basis equations. Thus, the possibility arises of including into the description the various waves identified in physical experiments. Currently, increasing attention is being paid to different types of wave interaction [1.15,39,40].

The first attempts to develop a universal approach to this problem were made by *Novikov* [1.41] and *Leble* [1.28]. The correct introduction of branch subspaces in the wave dispersion relation is necessary even in a linear theory. This is not a simple problem and is important for correct separation of the corresponding contributions in given initial data, a boundary regime or sources [1.42]. The approach of this book is based on a *projection operators* technique which is a convenient tool for deriving equations in their explicit form. This problem has an important physical aspect: the identification of the contribution of each wave type implies the measurement of a fundamental set of dynamical variables and the study of the connection between them (polarization relations). Therefore, the introduction of projection operators that define these relations is logical.

The modern nomenclature is based on linear equations. Therefore it is convenient to preserve the traditional terminology, at least in the case of a weak nonlinearity. The classification principle within the projection operators approach is general and thus allows generalization to the nonlinear case. In this book, however, only the linear couplings are given.

Let us explain the essence of the question by using simple examples. If in the nonlinear string equation

$$u_{tt} - c^2 u_{xx} + \sigma(uu_x)_x + \beta^2 u_{xxxx} = 0 \tag{1.3}$$

we substitute for u a sum of right Π and left Λ waves $u = \Pi + \Lambda$ and we require

$$\Lambda_t - c\Lambda_x = f_1 ,$$
$$\Pi_t + c\Pi_x = f_2 . \tag{1.4}$$

Then by calculating the sum and difference of equations (1.4) and differentiating the results with respect to t, x we get

$$u_{tt} - c^2 u_{xx} = (f_1 + f_2)_t - (f_2 - f_1)_x c . \tag{1.5}$$

It is clear that without additional assumptions the functions f_1 and f_2 cannot be defined. Such an additional assumption may be the condition requiring the absence of increasing contributions when Π and Λ are spatially separated at some interval $[0, t]$. By means of the D'Alambert formula it may be shown that

$$f_1 + f_2 = 0 . \tag{1.6}$$

From (1.5,6) we find $f_{1,2}$ as a class of functions which quickly decrease toward infinity and reduce to (1.4)

$$\Lambda_t - c\Lambda_x + \frac{\sigma}{c}(\Lambda + \Pi)_x^2 + \frac{\beta^2}{2c}(\Lambda + \Pi)_{xxx} = 0$$
$$\Pi_t + c\Pi_x - \frac{\sigma}{c}(\Lambda + \Pi)_x^2 - \frac{\beta^2}{2c}(\Lambda + \Pi)_{xxx} = 0 . \tag{1.7}$$

We see that the nonlinear terms that describe the interaction have been divided into two equal parts. This method is useful but does not allow for generalization.

The problem of the discrimination and interaction of waves is solved by the transition from the variables u, u_t defining the Cauchy problem to a new pair of variables $a = cu_x, b = u_t$ and the definition that the connection between them defines the right and left waves. To this end the system

$$b_t - ca_x = \sigma V(u) ,$$
$$a_t - cb_x = 0 ,$$

which is equivalent to the nonlinear Klein-Fock equation $u_{tt} - c^2 u_{xx} = \sigma V(u)$ is written in vector form

$$\psi_t + L\psi = N(\psi) \ . \tag{1.8}$$

Elementary calculations give the projection operators for the eigen subspaces of the operator L, which corresponds to the roots of dispersion equations $w = \pm k$

$$P_\pm = \frac{1}{2}\begin{pmatrix} 1 & \pm 1 \\ \pm 1 & 1 \end{pmatrix} \ . \tag{1.9}$$

Every solution of the linear problem now has a definite relation between the components of the vector ψ, namely, $\psi_+ = \{a,a\}, \psi_- = \{a,-a\}$. Applying the operators $P_\pm$ to (1.8) we have the system for right $\Pi = (a-b)/2$ and left $\Lambda = (a+b)/2$ waves

$$\Pi_t + c\Pi_x = -\frac{\sigma}{2}V\left(\frac{1}{c}\int_{-\infty}^{x}(\Pi + \Lambda)dx\right) = -\Lambda_t + c\Lambda_x \ .$$

The explicit form (1.3) of V gives (1.7) when $\Pi = \Pi_x$, $\Lambda = \Lambda_x$. The initial conditions for Π and Λ can be easily determined too.

The complexity of the wave separation problem depends on the number of dynamical variables and orders of derivatives in the equations of the fundamental problem.

The separation into right- and left-travelling waves may be obtained via a multiscale decomposition [1.4,23,43] as well as by the reductive perturbation method [1.44] as it is shown for the nonlinear string case. However, the separation of internal and sound waves in the atmosphere cannot be so simply treated. It is shown in Chap. 4 that with the discrete Silin–Tikhonchuk equations written in the form including interacting unidirectional Langmuir waves the system can be integrated exactly.

The rapid developments in the theory of integrable systems often lead to the proposal of new equations and their solutions before such equations are derived from physics. This was the case with the nonlocal Kadomtsev–Petviashvili equation analogue, the two-dimensional Joseph equation, which was investigated in [1.45] and derived in [1.27,46]. Nonlocal operators in the theory of electromagnetic wave dispersion are introduced when spatial dispersion or a strong inhomogeneity in the stratification in the medium of propagation exist [1.47] (Sect. 3.4). Such strong variations in the stratification scale of liquid media lead to nonlocal dispersion of internal waves too. In these cases the waveguide regions appear in a propagation medium where the solutions oscillate and exponentially decrease outside the medium. Then the basis of the linear problem is introduced inside the waveguide interval matching the solutions at the region's boundaries. The solution outside the guide interval may be found without mode expansion due to simplification of the structure of the basic equation operator. This problem was first studied at the level of a dispersion relation by *Phillips* [1.48] and *Whitham* [1.49]. The explicit form of the pseudo-differential operators appeared in the works of *Benjamin* [1.50], *Ono* [1.51] and *Levikov* [1.52] for the fluid pycnoclyne guide (layer with density varying in z) surrounded by infinite homogeneous layers in the single mode case. For a finite-depth fluid the nonlocal

KdV equation analogue has been derived by *Joseph* [1.53]. The two-dimensional multi-mode wave has also been studied [1.27]. The theory of nonlocal equations led to algebraic solitons [1.51]. The theory of algebraic and algebro-geometric integration was developed for nonlocal dispersion in [1.54].

The soliton solutions of single mode evolution equations are generally a rather crude approximation in waveguide theory. However, they can be the starting point for the averaging methods of *Whitham* [1.1,3]. The solution of the multi-mode problem is to look for a soliton form with parameters that weakly depend on the coordinates. Knowledge of the behavior of the solutions of simplified equations allows one to apply the iteration procedure to the system considered going over to the model equation in a power series of a small parameter (Sects. 2.4,5). The fundamental statement of the problem has been discussed by *Maslov* and *Dobrokhotov* [1.20–22] as well as in [1.55].

An important case of waveguide propagation is when only one mode is possible due to the waveguide dimensions. This is the case for electromagnetic waves in metal tubes, dielectric layers and fibers (Chap. 2). The broad range of possible applications in physics and engineering results in a prolifery of publications on model evolution equations of NS type with various kinds of higher order nonlinear dispersion and dissipation [1.12,56]. The total dispersion operator preserves the soliton form and the existence of multi-soliton decay gives the possibility of shortening the pulse time to femtoseconds [1.57].

The elimination of the arbitrariness in the choice of boundary conditions [1.26,56] requires an additional analysis of basic physical assumptions. For example, the transition from hydrodynamic flow to the collisionless regime in gaseous flow may be the cause of waveguide propagation [1.59]. The boundary conditions for such a wave propagation problem are formulated within the kinetic theory. The statement of the problem in the hydrodynamic regime follows from the general kinetic formulation.

There are many situations when the boundary regions along the longitudinal coordinates should be taken into account. In this case new salient features appear. For example, periodic conditions may arise in problems with a torus (ring) guide geometry, as is the case in experiments with rotating vessels [1.9,60] and tokamak plasma installations [1.39]. The fundamental mathematical elements of these problems are the Riemann θ-functions or finite-gap solutions [1.62]. The problem of extracting the class of real nonsingular periodic solutions from a general finite-gap solution has been discussed [1.62,63]. Large-scale atmospheric wave propagation, where the periodicity along the planet longitude is natural, allows for the application of the finite-gap integration method. For long internal waves the simplest two-dimensional single mode system is the KP equation [1.27,33]. The aspects of application of the θ-functional KP solution are specified in [1.64,65]; see also Sect. 2.5. The semi-infinite and closed interval solutions for integrable equations may be found in [1.66].

1.2 Overview

The second, fifth and sixth chapters of this monograph are devoted to fluid and gas dynamics. In Sect. 2.1 the fundamental differential mass, momentum and energy conservation equations as well as the equations of state are given in a form that is convenient for applications to atmospheric and ocean physics. The transition to the theory of incompressible fluids with dissipation is discussed. It is noted that the condition div $v = 0$ contradicts the energy equation with a nonzero thermoconductivity term. This point, which was neglected in [1.26], has corollaries in the mean field generation theories of Chap. 6. The aim of Chap. 2 is the identification and investigation of the contributions of various waves – acoustic, internal (gravity) and Rossby (planetary) – in a wave disturbance of a medium. This problem is formulated mathematically in Sect. 2.1. In addition, the hydrodynamical system of equations is transformed exactly to a system of two equations. This new system is approximately reduced to single equations in Sect. 2.4 and Appendix 1. These equations for special wave types with latitude dependent coefficients are necessary for the mode basis definition and further approximations. The method of this derivation differs from that of [1.2,30] and allows one to proceed without narrowing the wave range. In Sects.2.2,3 the basic principles of reducing the fundamental systems of equations are demonstrated for the classic problem of Rossby and Poincaré waves. The 3×3 projection operators for three dispersion branches are given in the explicit form here and in Appendix 2. The more complex case of 4×4 operators for subspaces of sound and internal waves is carried out in Appendix 3. The wave separation is done in the subspace of any definite transverse mode.

The wave interaction system is derived in the next section. This system contains long wave (KdV) as well as short wave nonlinear and dispersion terms. Detailed expressions are given for interaction constants in the long wave range. It is shown that nonlinear dispersion of long Rossby waves is described by a generalized CKdV equation [1.29,30].

Sections 2.4–6 are central to the chapter and the book. They present the idea of the space problem solution method in the mode representation using the separation of contributions with different dispersions. Section 2.4 contains a short review of internal wave theory and the derivation of interacting modes that are generalizations of KdV and KP equations. Dissipation, curvature and rotation of a planet are taken into account. We should emphasize that the KP case has a form which differs from the traditional [1.46], due to the interchange of x and t as in the magnetoacoustic plasma wave equation [1.67]. The forms are equivalent only if the horizontal velocity rotor component tends to zero [1.33]. The dissipation and dispersive effects in long internal wave propagation are discussed in connection with the hydrostatic approximation [1.68]. This approximation is equivalent to the NES with neglible KdV dispersion terms. It is shown that the presence of the main dissipation term in isolated mode equations does not stop the waves from inverting. It turns out that discontinuous solutions of these Riemanian equations with the dissipation term proportional to an unknown function also do not exist.

Therefore even the form of hydrostatic approximation waves radically differs from the dispersing waves at large times.

In Sect. 2.5, the perturbation parameter σ and valid time intervals are evaluated. The evaluation scheme has many common features with the approaches of *Maslov–Dobrokhotov–Omelyanov* [1.20–22,69] and *Taniuti–Wei* [1.16]. It is shown that advancing along the perturbation theory steps either widens the validity interval or refines the results in the fixed interval $t \in [0, \sigma^{-1}]$. Here the explicit solutions of the CKdV in first and second order in σ are presented. For illustration, the collisions of the stable multi-mode disturbances (quasisolitons) and the decay of the initial conditions are investigated. The analytic formulas of CKdV solutions have been tested by direct computer methods and analogous modelling of nonlinear coupled LC-circuits.

In the last section of the second chapter interactions of barotropic planetary waves are studied. The atmosphere is taken to be a toroidal waveguide, periodic along the longitude. Strongly coupled triads are extracted by larger interaction constants that are numerically calculated and tabulated with explicit formulas. The signs of the nonlinear terms of triads correspond to the nonlinear explosion instability. The corresponding periods are compared with experimental data [1.9]. Three-wave systems, in the particular case of two equal velocities, are transformed to the sine-Gordon (SG) equation just as in the theory of zonal flow and Rossby waves interaction [1.60].

The nonlinear equation system (NES) for the propagation of electromagnetic waves in a guide is considered in the third chapter. A similar approach to waveguide evolution equations is developed in [1.70,71]. Traditional guide problems acquire new characteristics under nonlinear conditions. The possibility arises of investigating mode generation, whose intensity and form depend on the properties of matter that constitutes the waveguide. In particular, it is clear that during a propagation process new (initially absent) modes may be created. In Sect. 3.1 the structure of linear guide modes is reproduced in a form that is convenient for applying nonlinear theory. The universality of the NS equation and its possible modifications [1.12,15,56] are discussed. The equations for interaction and the formulas for coupling constants are derived. The three-wave resonance conditions are investigated in Sect. 3.2 and Appendix 4. The special problem of the resonance range of electromagnetic waves of NMR-NQR is discussed for a nuclear spin system of dielectrics (Sect. 3.3), in which the interaction of different components of a magnetic field appears. The solution of the density matrix equation for the nuclear spin system (which may be directly repeated for any n-level pseudospin system) gives the nonlinear coupling constants as a function of transition frequencies and relaxation times. The three-wave system and its explicit solutions for experiments are derived by means of the reduction to SG or ShG equations. In Sect. 3.4 the propagation of an electromagnetic wave in a dielectric slab is considered. Its model equation contains an integral dispersion operator with a Hankel function core. The simple geometry allows one to show the boundary matching procedure in detail (see also Chap. 5 for generalizations).

Sections 4.1 and 4.2 are on the methodology and are addressed to mathematicians to facilitate the understanding of the physical content of Sects. 4.2–4. The theoretical foundation follows the approach of *Silin* [1.72] and [1.47]. In Sect. 4.2 the NES derivation is developed for inhomogeneous nonequilibrium plasma. The Langmuir and ion-acoustic wave interactions are considered to yield the Zakharov system of equations [1.73] with a new relaxation term due to plasma inhomogeneity. The three-wave helico-acoustic interaction is also studied. Section 4.3 is based on [1.37,74] and continues in the same vein. It contains the statement of the problem of the dynamics of parametric weak plasma turbulence. In Sect. 4.4 the theory of the algebraic integration of the discrete Silin–Tikhonchuk equations is given. These describe the interaction of antiparallel Langmuir waves [1.75]. The equations are a new example of an integrable system and contain solitons with new properties. By the theorems given in [1.28] the Darboux transform which generalizes the Matveev results [1.76] is chosen.

Chapter 5 summarizes the results of this author's work on quasi-waveguide propagation. Here a mode expansion is applied to a wave capture interval and its boundaries are defined by self-consistent conditions. In Sect. 5.1 the general technique, for any wave type, is outlined. In Sect. 5.2 the derivation of the mode system for internal pycnoclyne waves is shown. For a single mode the system yields the equation that unites the KdV, KP and Joseph-BO equations. These equations may be obtained as the limiting cases of the geometrical parameters of the pycnoclyne. The soliton solutions of those equations are compared to the results of laboratory experiments [1.8]. We note that the equation for the propagation of an internal wave in a quasi-waveguide, for example in the atmosphere, has a nonlocal dispersion term in the same form as that for an electromagnetic wave in a dielectric layer (compare Sects. 5.3 and 3.4).

An important direction of the theory is on wave propagation in media with varying Knudsen number (Kn) [1.77]. The results of sound propagation investigations at arbitrary Kn are reported in [1.78] and treated by a linear Boltzmann equation method for homogeneous media in [1.79]. Both theory and experiment demonstrate wave-like properties in Knudsen regimes that, however, strongly differ from hydrodynamical. The first attempts at a kinetic description of waves in stratified media were carried out in [1.80,81]. These attempts may be considered as "naive": the dispersion relations in [1.80] coincide with hydrodynamical ones and the applicability of the work reported in [1.81] is restricted to the upper layers of the atmosphere; there the regime is considered to be collisionless, with collision effects taken as a perturbation increasing with decreasing height. The awkward formulas of the latter approach do not allow one to move into the Knudsen flow regime effectively.

The approach of the Sect. 5.4 is based on matching the distribution functions across a transition boundary layer where the distribution function is a power series in the transverse coordinate. This method is similar to the Fuchs models [1.82] where the problem of vapor condensation is solved in the zero transition layer approximation by direct patching of collisionless and hydrodynamic descriptions. The similar "zero" version of a matching theory for acoustic-gravity waves gives the upper boundary condition for a hydrodynamic atmosphere waveguide that closes the description in it. In the local equilibrium region, the usual results with the new wavenumber spectrum are realized. In the Kn $\gg$ 1 region the deviation from ideal hydrodynamics is radical: instead of exponential growth, exponential decrease of a wave amplitude occurs.

In Chap. 6 the dynamics of a medium – wave system are studied. In Sect. 6.1 the dynamical description of a medium is introduced and the dynamical equations that determine the interaction between a shear flow and internal waves are derived (see also Appendix 5). Large-scale disturbances that appear as a self-organized wave are called the mean field (motion) by *Grimshaw* [1.83] and *Miropolsky* [1.26]. It is shown that the Zakharov system for plasma waves [1.73] is general for interactions of both short and long waves (that may be related to mean field). Investigations in nonlinear internal wave evolution in mean fields generation shows the substantial contribution of modes with high numbers [1.26,73]. Accounting for the rapid oscillation in the transverse (vertical) coordinate necessitates the inclusion of dissipation effects. The influence of thermoconductivity and viscosity is considered in Sect. 6.2 with a determination of the transport coefficients including background turbulence [1.84]. As a result of the internal wave action a field with a small vertical scale is generated. This mechanism gives an explanation of the fine pycnoclyne structure that frequently occurs in the ocean [1.85]. An instability of a temperature field caused by an internal wave in the presence of dissipation is considered in Sect.6.4. Nonsingular perturbation theory in dissipation problems of mean field interactions of internal waves gives the resonant generation of the small-scale terms that are described by the modified Zakharov equations. The modified incompressibility condition is taken into account (Sect. 6.5).

In Sect. 6.3 the analytic theory of ionospheric plasma response to a long internal wave is discussed. The parameters of the internal wave are included as the coefficients in the plasma ambipolar diffusion equation. The perturbation scheme allows one to get the explicit formulas for plasma density and maximum position variations in the F-layer of the ionosphere in terms of Whittacker functions. Details, as well as a comparison with observations of disturbances caused by the auroral current are given in [1.65].

Before concluding, we would like to point out one more application of the solutions for various dispersion branches in numerical simulations of complex systems of equations. The projection of a finite-difference scheme onto the subsolutions of long or short scales at every step of the calculation procedure may preclude the generation of the additional wave contribution. This may give rise to error accumulation. Thus the projection representation for dynamical variables provides a convenient language for the mathematical statement of the problems and for the physical interpretation of observations.

2. The Discrimination and Interaction of Hydrodynamical Waves of Different Types

In this chapter the main features of hydrodynamic waves are described and the basic ideas of this book are introduced with examples taken from hydrodynamics and gas dynamics. The fundamental notions of elementary interacting modes and quasisolitons (approximate stable solutions) for long waves are discussed in the framework of the nonsingular perturbation theory. The corresponding small parameters are defined in terms of initial (boundary) conditions (Sects. 2.1–4) and some simple estimations of perturbation terms are given in Sect. 2.5. The form of the projection operators to the subspaces of elementary modes is given in Sect. 2.3. The theory of the coupled KdV systems as a basic tool for long waves in waveguides is developed in Sects. 2.5–6. The special case of closed waveguide propagation of Rossby waves is analyzed in Sect. 2.7.

2.1 Hydrodynamic Wave Equations in a Three-Dimensional Stratified Medium

2.1.1 Fundamentals

For the sake of completeness we begin our discussion with the physical foundations of the theory of wave motion in an ideal fluid and dissipation effects. In Sect. 5.4 the role of equations is discussed and in Sect. 6.2 the effects of turbulence are considered. Stationary states that do not affect the dynamical variables are also introduced.

In a nonlinear theory the determination of the ground state is nontrivial. The assumption of stationarity is to some degree arbitrary and requires justification. Indeed, in experiments, large scale motion is regarded as a background phenomenon but the mutual interaction of such motions may give rise to nonlinear effects of infinite intensity. Therefore, in this book we shall refer to a ground state only when we mean an elementary stationary state without convective nonlinearities. All other states we shall initially consider to be dynamically interacting (Sect. 6.1).

Traditionally we denote by z the vertical and by x, y the horizontal coordinates. We introduce the the mass density of the medium $\varrho = m/V$ where m is the mass in the volume V. By the vector v we denote the mass velocity with the components $v_1 = v_x = u$, $v_2 = v_y = v$, $v_3 = v_z = w$. If the medium is complex and can be described by the Boltzmann equations, then introducing the different components of the density $\varrho_i = m_i/V$ and the mean velocities v_i we can write

$$\partial \varrho_i / \partial t + \text{div}\, \varrho_i(v + V_i) = 0 \tag{2.1}$$

where V_i is the diffusion velocity that is determined from the mean thermal velocity $v_i - v$, $v = \sum_i \varrho_i v_i / \varrho$. Equation (2.1) is the basis of diffusion theory (Sects. 6.2,3). After summing over the particle index i in (2.1) we get

$$\varrho_t + \text{div}\, \varrho v = \frac{d\varrho}{dt} + \varrho\, \text{div}\, v = 0 \tag{2.2}$$

since $\sum_i \varrho_i V_i = 0$. Equation (2.2) is the hydrodynamical mass flux continuity equation. We refer the reader to [2.1–3] for the details of the derivation using kinetic and thermodynamical approaches to the energy equation and give here only a brief outline.

The differential of the internal energy e is a function of the pressure p and the heat transferred δQ: $de = -pdV + \delta Q$, by differentiating the relation $\varrho = \mu/V$ with respect to t: $dV/dt = -(\mu/\varrho^2)(d\varrho/dt)$ and using (2.2) we get for a mole with mass μ

$$\frac{de}{dt} = -\mu \frac{p}{\varrho} \text{div}\, v + \frac{\delta Q}{dt}. \tag{2.3}$$

Both phenomenological [2.4] and kinetic considerations [2.1] give for a simple medium

$$\frac{\delta Q}{dt} = c_v \, \text{div}\, \kappa \nabla T + \frac{\mu}{\varrho} q, \tag{2.4}$$

where κ is the coefficient of thermoconductivity, T is temperature, q is the provided power per unit volume. This equation is approximately valid for a many-component medium if the mass fluxes are small [2.1].

The equation of motion for a viscous simple medium in a rotating frame with angular velocity Ω is

$$\varrho \frac{dv_i}{dt} + 2\varrho[\Omega \times v]_i = -\nabla_i p + \nabla_k \sigma_{ik} + F_i,$$

$$\sigma_{ik} = \eta \left(\nabla_k v_i + \nabla_i v_k - \frac{2}{3}\delta_{ik}\, \text{div}\, v\right) \tag{2.5}$$

where η is the dynamical viscosity coefficient and F_i is the volume density of the force field. For example, in a gravitational field $F = \varrho g$, where g is the acceleration due to gravity, $\nabla_i = \partial/\partial x_i$. The dissipative coefficients κ and η are phenomenological. The system is closed by the equations of state

$$\varrho = \varrho(p, T), \quad e = e(p, T). \tag{2.6}$$

In atmospheric physics (2.6) are known as the Clapeyron–Mendeleev and the energy equation [2.4], and are written

$$\mu p = \varrho RT , \quad e = c_v T , \tag{2.7}$$

where c_v is the molar heat capacity at constant volume. Putting (2.7) into (2.5) we exclude pressure and energy

$$c_v \frac{dT}{dt} + RT \operatorname{div} \boldsymbol{v} = c_v \operatorname{div} \kappa \nabla T + q \frac{\mu}{\varrho} ,$$

$$\varrho \frac{dv_i}{dt} + 2\varrho [\boldsymbol{\Omega} \times \boldsymbol{v}]_i + \frac{R}{\mu} \nabla_i \varrho T = \nabla_k \sigma_{ik} + \varrho g_i . \tag{2.8}$$

For water we can adopt the linearized equations of state (2.6)

$$\varrho = \varrho_0 (1 - \nu \tau) + \beta \varrho_0 (p - p_0) , \tag{2.9}$$

$$e = e_0 + c_p \tau . \tag{2.10}$$

The temperature τ is measured from the level at which $\varrho = \varrho_0$, $e = e_0$, $p = p_0$. In the ocean, typically, $\nu \sim 10^{-4} K^{-1}$, $\beta = 10^{-10} \text{Pa}^{-1}$. When describing internal waves we do not take into account the last term in (2.9). The equations of conservation of energy and momentum are then

$$c_p \frac{d\tau}{dt} = -p \operatorname{div} \boldsymbol{v} + c_p \operatorname{div} \kappa \nabla \tau + q , \tag{2.11}$$

$$\varrho \frac{dv_i}{dt} + 2\varrho [\boldsymbol{\Omega} \times \boldsymbol{v}]_i = -\nabla_i p + \nabla_k \sigma_{ik} + \varrho g_i . \tag{2.12}$$

In this book we are interested in those problems where dissipative effects can be treated via perturbation theory. We assume that the initial conditions are smooth enough and that the nonlinearity is weak and does not lead to large gradients. Special cases of this situation are mean field generation (Sects. 6.4, 6.5) and the increasing kinematic dissipation (Sect. 5.4). The role of thermoconductivity and viscosity is discussed in particular for these cases. Here, however, we shall first investigate the linear nondissipative problem and then introduce nonlinearity and dissipation effects by means of the nonsingular perturbation theory. For specific conditions we shall take into consideration the degree of background and introduced turbulence that influence the transport coefficients [2.5].

We choose our ground state to be stationary and denote the corresponding parameters by a bar. The wave variables are denoted by primes. For example, $T = \bar{T} + T'$. In the stationary state $\boldsymbol{v} = 0$ and therefore from (2.5) it follows that

$$-\nabla \bar{p} + F_0 = 0 , \tag{2.13}$$

$$\bar{\varrho} c_v \operatorname{div} \kappa \nabla \bar{T} + q_0 \mu = 0 , \tag{2.14}$$

where F_0 and q_0 are the force and power densities that do not contain the sources of wave disturbances. In a medium stratified along z it is usually assumed that $\bar{p}$ and $\bar{T}$ depend only on z. Sometimes a weak dependence on x of some small parameter is allowed. Thus, $\bar{p}_z = F_{0z}$, $(\kappa \bar{T}_z)_z = -q_0(\mu/\bar{\varrho} c_v)$. If $F_{0z} = g\bar{\varrho}(z)$ then

$$\bar{p} = g \int_0^z \bar{\varrho}(z)dz + p_0, \quad \bar{T} = -\int_0^z \kappa^{-1}\mu \int_0^{z'} \frac{q_0(z'')}{\bar{\varrho} c_v} dz'' dz' + T_0 + T_1 z$$

where $\bar{p}(0) = p_0$, $\bar{T}(0) = T_0$, $\bar{T}_z(0) = T_1$. The equations for the wave are

$$\frac{d\varrho'}{dt} + w\bar{\varrho}_z + (\bar{\varrho} + \varrho')\,\text{div}\,\boldsymbol{v} = 0, \tag{2.15}$$

$$\frac{de'}{dt} + w\bar{e}_z + \mu\frac{\bar{p} + p'}{\bar{\varrho} + \varrho'}\,\text{div}\,\boldsymbol{v} = c_v\,\text{div}\,\kappa\nabla T' + q'\frac{\mu}{\bar{\varrho}}, \tag{2.16}$$

$$\varrho\left(\frac{dv_i}{dt} + 2[\boldsymbol{\Omega}\times\boldsymbol{v}]_i\right) = -\nabla_i p' + \nabla_k \sigma_{ik} + F'_k. \tag{2.17}$$

In the case of Cauchy problem without external sources $q' = 0$, $\bar{F}' = \varrho' \bar{g}$.

2.1.2 Atmospheric Waves: Dispersion and Polarization

The study of nonlinear processes requires a knowledge of the characteristic space and amplitude scales. Therefore it is necessary to specify boundary and initial conditions. Let us consider the mathematically and practically fundamental example of an atmosphere whose density varies exponentially with distance.

We assume that the background temperature depends weakly on the altitude. Then the density of the atmosphere decreases exponentially with the height z by

$$\bar{\varrho} = \varrho_0 \exp(-z/H) \tag{2.18}$$

where ϱ_0 is the density at the earth's surface and H is a "height scale" or a "homogeneous atmospheric height" that is proportional to the background temperature $\bar{T}$; $H = R\bar{T}/(\mu g)$ [2.4]. This approximation holds in atmospheric waveguides over intervals of z where the temperature gradients $\bar{T}_z$ are small. These intervals correspond to layers, at the boundaries of which the properties vary sharply (Sect. 5.3) [2.4,6].

Let us write the equations (2.15–17) neglecting dissipation and accounting for (2.18) and (2.8). We also introduce the variable $\delta' = H\ln(\varrho/\bar{\varrho})$. The new dynamical function δ' for small perturbations of the medium corresponds to the rise of a gas (fluid) volume element which oscillates when a wave passes by.

To simplify the problem from three parameters R, H, g with different dimensions, we convert to dimensionless variables and their derivatives. We shall denote the variables for δ', T' by the same letters but without primes

$$\delta' = H\delta, \quad T' = \bar{T}T = \mu g H T/R.$$

The designations for the other variables will not be changed. It is sufficient in (2.15–17) to put $\boldsymbol{r} \to H\boldsymbol{r}$, $\boldsymbol{v} \to \boldsymbol{v}\sqrt{gH}$, $t \to t\sqrt{H/g}$. Collecting the linear terms in the left part we get

$$\delta_t + \text{div}\,\boldsymbol{v} - w = a, \tag{2.19}$$

$$v_t + 2[\Omega \times v] - Tn + \nabla(\delta + T) = b \,, \tag{2.20}$$

$$T_t + (\gamma - 1)\,\mathrm{div}\,v = d \,. \tag{2.21}$$

a, b, d contain both nonlinear and dissipative terms as well as possible sources. Note that $\gamma = c_p/c_v$, $n = r/r$.

The statement of the problem and therefore the choice of a solution method for (2.23–25) depends on the wave type and the frequency range. The wave type is determined by the dispersion relation and the relationship between dynamical variables u, v, w, δ, T (Chap. 1) [2.7]. The latter are especially important in nonlinear theory and allow to determine unambiguously the form of the interaction operator. Thus, it is known that in the atmosphere (and in the ocean) there exist three classes of waves: the longest period Rossby waves ($\omega < \Omega$), internal waves $\omega < \sqrt{g(\gamma-1)/(H\gamma)}$ and acoustic waves $\omega > \sqrt{g\gamma/(2H)}$. The internal and acoustic waves correspond to two roots of the dispersion equation (dispersion branch) which are different in sign. Therefore, there are actually five types of waves which can be determined by the order of time derivative in (2.19–21) [2.6,8].

If the excited waves are short compared to height H then during the time of the propagation inside the atmospheric waveguide the Cauchy problem can be solved by the ray approximation method [2.9]. If the dimensions of the wave are of the order of the waveguide dimensions, then a mode representation with discrete vertical wavenumber is used. When the horizontal wavelength is of the order of the planet radius then the solution must include a condition of periodicity along the longitudinal axis φ and of finiteness in the interval of possible variations of colatitude ϑ. Then the horizontal wavelength and frequency spectrum become discrete (Sect. 2.7).

The lower boundary of the atmosphere, i.e., the planet surface, imposes an obvious boundary condition $v_n = 0$, but the condition on the upper boundary is not as straightforward to determine [2.10–12]. In Sect. 5.4 we discuss the upper boundary condition on the vertical mass velocity projection. (This is, because of $\gamma w + w_z = 0$, a boundary condition of the third kind; in contrast to $w_z = 0$ and $w = 0$ corresponding to the second and first kind, respectively.)

Thus, the classification of the wave disturbances includes the following steps:
(1) the derivation of dispersion equation;
(2) analysis of the roots of the equation;
(3) construction of the projection operators for the subspaces of dynamical variables of the roots of the dispersion equation.

It should be emphasized that points (2) and (3) can be considered as the physical determination of the wave type (Chap. 1, Sect. 2.2). It is by the connection between the "wave vector function" components that we can identify the oscillations in the case of overlap of the frequency ranges. The knowledge of the projection operators allows one to single out the initial condition subspaces that generate a wave of a particular type (sound, internal gravity wave, Rossby wave). In the nonlinear problem, waves of different types effectively interact [2.13] and such projection operators enable one to introduce correctly the "interaction Hamiltonian" according to the physical determination [2.7,14].

The notion of the dispersion equation in an inhomogeneous medium must be made more specific. If the wavelength λ of a disturbance is smaller than the minimal scale of inhomogeneity H then the local dispersion equation can be used within the framework of asymptotic methods with the small parameter λ/H. If such a condition is not valid then one should combine the mode expansion with the construction of the dispersion relation for each mode [2.14–16]. In this case of long waves with wavelength on the order of the earth radius, it is necessary to investigate all spectral parameter relations and determine the frequency dependencies by them (Sect. 2.7) [2.17]. In some problems, e.g., exponential atmospheric waves over a planar earth surface, the fundamental system can be reduced by substitution to equations with constant coefficients. Then the usual dispersion analysis may be applied. We shall continue the investigation of the exponential atmosphere over a rotating surface.

Let us decrease the number of unknown functions in the system (2.19–21). The vertical component of the velocity is excluded from the combination of (2.19,21):

$$w = \delta_t - T_t/(\gamma - 1) - a + d/(\gamma - 1) ,$$

and

$$\operatorname{div} \boldsymbol{v} = d/(\gamma - 1) - T_t/(\gamma - 1) . \tag{2.22}$$

Calculating the rot and div of the equation of motion (2.20) and using vector algebra we get

$$\operatorname{div} \boldsymbol{v}_t - 2(\boldsymbol{\Omega}, \operatorname{rot} \boldsymbol{v}) - T_z + \Delta(\delta + T) = \operatorname{div} \boldsymbol{b} , \tag{2.23}$$

$$\operatorname{rot} \boldsymbol{v}_t + 2\boldsymbol{\Omega} \operatorname{div} \boldsymbol{v} - 2(\boldsymbol{\Omega}, \nabla)\boldsymbol{v} - [\nabla T \times \boldsymbol{n}] = \operatorname{rot} \boldsymbol{b} . \tag{2.24}$$

We introduce $[\boldsymbol{\Omega} \times \boldsymbol{n}] = \boldsymbol{m}$ to get the oblique coordinate system based on $\boldsymbol{m}$, $\boldsymbol{n}$, $\boldsymbol{\Omega}$. After scalar multiplication of (2.24) by $\boldsymbol{\Omega}$ we differentiate the result by t using (2.21,22). We put (2.22) into (2.23) and differentiate again by t, expressing rot $\boldsymbol{v}_t$ by the transformed (2.24). This gives

$$\begin{aligned}
&-(T_{tt} + 4\Omega^2 T)_{tt}/(\gamma - 1) - 4(\boldsymbol{\Omega}, \nabla)[T(\boldsymbol{\Omega}, \boldsymbol{n}) - (\boldsymbol{\Omega}, \nabla)T] \\
&+ 2(\boldsymbol{m}, \nabla T_t) - (\nabla, \vec{n})T_{tt} + \Delta T_{tt} + \Delta \delta_{tt} + 4(\boldsymbol{\Omega}, \nabla)^2 \delta \\
&= \operatorname{div} \boldsymbol{b}_{tt} - (d_{tt} + 4\Omega^2 d)_t/(\gamma - 1) + 4(\boldsymbol{\Omega}, \nabla)(\boldsymbol{\Omega}, \boldsymbol{b}) \\
&+ 2(\boldsymbol{\Omega}, \operatorname{rot} \boldsymbol{b}_t) \equiv F .
\end{aligned} \tag{2.25}$$

The second equation for δ and T can be obtained by projecting (2.20) onto the vertical direction $\boldsymbol{n} = \boldsymbol{r}/r$ and $\boldsymbol{m}$, $\boldsymbol{\Omega}$:

$$w_t + 2(\boldsymbol{n}, \boldsymbol{\Omega} \times \boldsymbol{v}) + (\delta + T)_r - T = (\boldsymbol{n}, \boldsymbol{b}) ,$$

$$(\boldsymbol{m}, \boldsymbol{v}_t) + 2\Omega^2 w - 2(\boldsymbol{\Omega}, \boldsymbol{v})(\boldsymbol{\Omega}, \boldsymbol{n}) + (\boldsymbol{m}, \nabla)(T + \delta) = (\boldsymbol{m}, \boldsymbol{b}) ,$$

$$(\boldsymbol{\Omega}, \boldsymbol{v}_t) = T(\boldsymbol{\Omega}, \boldsymbol{n}) - (\boldsymbol{\Omega}, \nabla)(\delta + T) + (\boldsymbol{\Omega}, \boldsymbol{b}) .$$

Dropping w and the vector and scalar products of Ω and v we come to

$$(\delta_{tt} + 4\Omega^2 \delta)_{tt} - (T_{tt} + 4\Omega^2 T)_{tt}/(\gamma - 1)$$
$$- 4(\Omega, n)^2 T + 4(\Omega, n)(\Omega, \nabla)T + 2(m, \nabla)T_t + T_{rtt} - T_{tt}$$
$$+ 2(m, \nabla)\delta_t + \delta_{rtt} + 4(\Omega, n)(\Omega, \nabla)\delta = (b_{tt}, n) + 4(\Omega, b)(\Omega, n)$$
$$+ a_{ttt} - d_{ttt}/(\gamma - 1) - 4\Omega^2 d_t/(\gamma - 1) + 4\Omega^2 a_t + 2(m, b_t) \equiv G \ . \quad (2.26)$$

The system (2.25,26) describes arbitrary waves in an exponential atmosphere and allows one to generate various approximations for the individual types of waves. For example, in the case of $\Omega = 0$ we get the equation for T which describes internal waves [2.18]. These equations should be used for the derivation of the equation for internal waves that take into account the earth rotation (Appendix 1).

We use the system (2.25,26) for the derivation of the equations for internal and acoustical waves taking into account the curvature of the earth and Coriolis effects. We rewrite (A1.3) in the form

$$T_{tttt} - \gamma \Delta T_{tt} + 2(\gamma - 1)\frac{T_{tt}}{r} - (\gamma - 1)\Delta_\perp T + \gamma T_{rtt}$$
$$- 4\Omega^2 \cos^2 \vartheta (T_{rr} - T_r) + \frac{2}{r}(2 - \gamma)\Omega T_{t\varphi} + 2(\gamma - 1)\Omega \int_0^t \frac{T_\varphi dt}{r^2}$$
$$= (\gamma - 1)F \ , \quad (2.27)$$

where F contains nonlinear and dissipative terms as well as source functions in the combination of (A1.2) Then we transform the left-hand side operator in spherical coordinates by expanding the coefficients of (2.27) in a Taylor series about $\vartheta = \vartheta_0$. Introducing local cartesian coordinates $z = r - a$, $y = a\varphi \sin \vartheta_0$, $x = a(\vartheta - \vartheta_0)$, where a is the earth radius and $f = 2\Omega \cos \vartheta_0$, we get

$$T_{tttt} - \gamma(\Delta T - T_z)_{tt} - (\gamma - 1)\Delta_\perp T - f^2(T_{zz} - T_z)$$
$$+ 2(2 - \gamma)\Omega \sin \vartheta_0 T_{yt} + \frac{2\gamma}{a}(\gamma - 1)\Omega \sin \vartheta_0 \int_0^t T_y dt$$
$$= F(\gamma - 1) \ . \quad (2.28)$$

In (2.28) we neglect terms of order h/a, i.e., the ratio of the depth of the atmospheric waveguide to the earth radius.

2.2 Projection Operators on Subspaces of Waves

2.2.1 Determination of Wave Type

For the sake of mathematical simplicity it is useful to adopt the idea of a dispersion relation branch – the subspace of solutions in which the time dependence is determined by the explicit dependence of the frequency on the wave vector components. Sound, internal and Rossby waves are related to such classes of waves in geophysical dynamics, as well as Kelvin, Lamb and Poincaré waves [2.4,8].

In the case of electromagnetic waves, either the components of the field intensity or the states of definite polarization can be similarly described [2.19]. There are also various types of plasma waves [2.20]. The complete determination of a wave type involves defining the system of dynamical variables – the physical equipment that gives a full description of the wave. Determination of wave type according to its frequency is not sufficient for complete identification because the frequency (wavelength) intervals can overlap.

Every measurement system used for identification can be correlated (as in quantum mechanics) with the projection operator on the subspace of a given dispersion branch. Mathematical description of this operator follows naturally from the language used in dispersion relations, i.e., from a linearized system of equations. The solution of the linear system has been elaborated and the system of eigenvectors for the matrix corresponding formulation is complete. The relations between the components of the eigenvectors are treated as polarization equations which need to be found. It is they that can give a physical determination of a wave type. It is this approach in the nonlinear theory which allows one to give an unambiguous form for the interaction (Chap. 1) [2.14,16].

The process of nonlinear evolution can be interpreted as the dynamics of interaction of waves of different types. The only restriction on the suggested wave classification system is that the measurement time must be smaller than the time of accumulation of nonlinear effects. It should be noted that projection operators are useful in the linear theory as well. They allow one to distinguish the relations between dynamical variables of a given wave type under the initial conditions in the simplest form.

2.2.2 Projection Operators in Geophysical Hydrodynamics

Let us explain the technique on examples taken from geophysical hydrodynamics. The classical example is the scheme leading to an interaction system of monodirectional waves in a one-dimensional (Chap. 1) or two-dimensional problem [2.7]. The waves in this case correspond to the opposite roots of the dispersion relation. In particular, a long surface wave is described by the KdV equation which is obtained from the Boussinesq system after projection on the subspace of a directed wave. The discrimination of directed internal waves is described in [2.7]. A more complicated example is given in [2.14]. In this latter paper the interaction of acoustic and internal waves is considered. Four projection operators similar to the ones constructed there are given in Appendix 3. Below we shall discuss a general scheme for construction of those operators for an exponential atmosphere over a rotating planet surface. In Sect. 2.3 we describe such an operator set for surface Poincaré and Rossby waves.

The system of equations (2.19–21) can be written in matrix form

$$\psi_t + L\psi = N(\psi) = -\psi^+ \otimes \hat{N}\psi \tag{2.29}$$

where $\psi^+ = (\delta, u, v, w, T)$ is a vector with the components describing the state of the atmospheric gas. The right-hand side of (2.29) can be given in tensor form

$$\{\psi^+ \otimes \hat{N}\psi\}_i = \psi^+ \hat{N}_i \psi \equiv \psi^+_\mu N_i^{\mu\alpha} \psi_\alpha,$$

which is convenient for further computations. The hydrodynamics equations (2.11,12) may be written in the same form as (2.29) too. The equation system for an electromagnetic field in a medium (Sect. 3.1) and for plasma (Sect. 4.1) may be arranged in the same way. Let us give the operators $L(\nabla)$ and $N(\nabla)$ for a nondissipative case:

$$L = \begin{pmatrix} 0 & \partial/\partial x & \partial/\partial y & \partial/\partial z - 1 & 0 \\ \partial/\partial x & 0 & -2\Omega_z & 2\Omega_y & \partial/\partial x \\ \partial/\partial y & 2\Omega_z & 0 & -2\Omega_x & \partial/\partial y \\ \partial/\partial z & -2\Omega_y & 2\Omega_x & 0 & \partial/\partial z - 1 \\ 0 & (\gamma-1)\partial/\partial x & (\gamma-1)\partial/\partial y & (\gamma-1)\partial/\partial z & 0 \end{pmatrix}$$

$$\hat{N}_1 = \begin{pmatrix} 0 & 0 & 0 & 0 & 0 \\ \partial/\partial x & 0 & 0 & 0 & 0 \\ \partial/\partial y & 0 & 0 & 0 & 0 \\ \partial/\partial z & 0 & 0 & 0 & 0 \\ 0 & 0 & 0 & 0 & 0 \end{pmatrix}, \quad \hat{N}_2 = \begin{pmatrix} 0 & 0 & 0 & 0 & 0 \\ 0 & \partial/\partial x & 0 & 0 & 0 \\ 0 & \partial/\partial y & 0 & 0 & 0 \\ 0 & \partial/\partial z & 0 & 0 & 0 \\ \partial/\partial x & 0 & 0 & 0 & 0 \end{pmatrix},$$

$$\hat{N}_3 = \begin{pmatrix} 0 & 0 & 0 & 0 & 0 \\ 0 & 0 & \partial/\partial x & 0 & 0 \\ 0 & 0 & \partial/\partial y & 0 & 0 \\ 0 & 0 & \partial/\partial z & 0 & 0 \\ \partial/\partial y & 0 & 0 & 0 & 0 \end{pmatrix}, \quad \hat{N}_4 = \begin{pmatrix} 0 & 0 & 0 & 0 & 0 \\ 0 & 0 & 0 & \partial/\partial x & 0 \\ 0 & 0 & 0 & \partial/\partial y & 0 \\ 0 & 0 & 0 & \partial/\partial z & 0 \\ \partial/\partial z & 0 & 0 & 0 & 0 \end{pmatrix},$$

$$\hat{N}_5 = \frac{1}{\gamma - 1} \begin{pmatrix} 0 & 0 & 0 & 0 & 0 \\ 0 & 0 & 0 & 0 & \partial/\partial x \\ 0 & 0 & 0 & 0 & \partial/\partial y \\ 0 & 0 & 0 & 0 & \partial/\partial z \\ 0 & (\gamma-1)\partial/\partial x & (\gamma-1)\partial/\partial y & (\gamma-1)\partial/\partial z & 0 \end{pmatrix}.$$

If we consider the wavelengths that are smaller than the radius of the earth, then the β-plane approximation and corresponding dispersion relation are valid. The dispersion equation can be derived via a Fourier transformation in x, y, z and represents the characteristic equation of a system of linear differential equations in t:

$$\det[i\omega I - L(i\mathbf{k})] = 0 . \tag{2.30}$$

The vector $\Omega(\vartheta)$ is taken at the colatitude ϑ_0 in the neighborhood of which the problem of propagation of the system of waves is solved. One can assure himself that (2.30) coincides with the dispersion relation taken from (2.28).

The equation

$$L(i\mathbf{k})\tilde{\psi} = i\omega\tilde{\psi}$$

in the Fourier transform space has generally five different eigenvectors and eigenvalues that depend on $\mathbf{k}$ [see e.g. Ref. 2.4]. Moreover, five projection operators $P^{(\alpha)}$ can be constructed for them: $(P^{(\alpha)})^2 = P^{(\alpha)}$, $P^{(\alpha)}P^{(\beta)} = 0$, $\sum_\alpha P^{(\alpha)} = I$. Examples of such operators are discussed above. The explicit forms for the eigenvectors and operators $P^{(\alpha)}$ in a general five-component formulation are not given here because of their complexity. The technique of their construction becomes clear from the examples.

The system of nonlinear equations for different types of interacting waves is obtained if $\psi = \sum_\alpha P^{(\alpha)}\psi$ is put into (2.29) and after $P^{(m)}$ is multiplied by the same equation

$$\psi_t^{(m)} + L\psi^{(m)} = \left(\sum_s \psi^{(s)}\right)^+ \otimes \hat{N}\left(\sum_q \psi^{(q)}\right), \tag{2.31}$$

where $P^{(m)}\psi = \psi^{(m)}$. The operators $P^{(m)}$ commute with L by definition. If the matrix coefficients in (2.29) depend on $\mathbf{r}$, a similar procedure can be developed when the variables in the linearized (2.29) are shared at least approximately [2.7]. This will be the case if the problem is (approximately) symmetric. A stratified medium is, in fact, an example of such a case. To a degree it has the symmetry of homogeneous media [2.15,21]. The rotation of the earth breaks the symmetry, however, but if the parameter β is small the deviation from total symmetry is small too. Then (Chap. 5) we start from the mode expansion in the linear problem and, afterwards, for each projection of the system (2.29) on every mode subspace we use (2.30,31).

2.2.3 Projection Operators for Poincaré and Rossby Waves

As mentioned earlier, the simplest examples of projection operators for right and left waves are given in Chap. 1. Let us consider another nontrivial example where the dispersion equation has three different roots, and the matrices of the projection operators are of the third order. An example of this is the classical problem of surface waves in a channel with a sloping bottom over a rotating planet. This problem corresponds to experiments with rotating tanks [2.22]. The theoretical interpretation for these experiments by means of nonlinear wave theory is given in [2.17]. Note that the sloping bottom can be a model for curvature of a planet surface. Thus, after proper modification of the statement of the problem, ocean and atmospheric waves on a planetary scale can be described in the framework of the calculated results. (The linear theory of the Poincaré and Rossby waves is brilliantly expounded in the book by J.Pedlosky [2.8].)

The system of equations for surface waves in a channel with an inclined bottom with the depth coordinate dependence $H = H_0(1 - \beta y)$ for the projections u, v of the hydrodynamical velocity on the axis x, y and η, the height of the surface over the horizon $z = 0$, and the assumption $uH_0 = U$ and $vH_0 = V$, $gH_0 = c_0^2$ is

$$U_t - fV + c_0^2 \eta_x = A = -(UU_x + VU_y)/H_0 ,$$
$$V_t + fU + c_0^2 \eta_y = B = -(UV_x + VV_y)/H_0 ,$$
$$\eta_t + U_x + V_y - \beta V = C = -(U\eta)_x/H_0 + \beta y(U_x + V_y) . \quad (2.32)$$

The boundary conditions are now conditions of impenetrability of the channel walls $V|_{y=0,1} = 0$. Let $\boldsymbol{k} = \{k, l\}$, then the dispersion relation can be written in the traditional form [2.8]

$$\sigma^2 - f^2 - c_0^2[\boldsymbol{k}^2 + \beta^2/4] - kc_0^2 \beta f / \sigma = 0 . \quad (2.33)$$

The roots can be found approximately. The first lies in the region $\sigma^2/f^2 \ll 1$ and corresponds to the Rossby wave

$$\sigma_1 = -\beta f k/(\boldsymbol{k}^2 + f^2/c_0^2) . \quad (2.34)$$

The rest determine the Poincaré waves

$$\sigma_{2,3} = \pm c_0 \sqrt{(f/c_0)^2 + \boldsymbol{k}^2 + \beta^2/4} . \quad (2.35)$$

Analyzing the equations relating U, V, η and the boundary conditions for V we come to a general form for the solutions of the system (2.32) with $A = B = C = 0$

$$V = \sum_n Y_n \theta_n ,$$
$$U = \sum_n (\varphi_n Y_n + \phi_n Y_{ny}) ,$$
$$\eta = \sum_n (\mu_n Y_n + \nu_n Y_{ny}) , \quad (2.36)$$

where $Y_n = (\sin l_n y)e^{\beta y/2}$ and $l_n = n\pi$. We can get the system for the expansion coefficients and write it without indices. Using algebraic identities from the system we exclude ϕ, ν. The system will be similar to (2.29). Denoting $Q = (l^2 + \beta^2/4)/(\beta^2 - f^2/c_0^2)$ and doing the Fourier transformation over x, $\theta = \int \exp(ikx)\tilde{\theta}dk$, we have

$$\tilde{\mu}_t + ik\tilde{\varphi} - \beta\tilde{\theta} = 0 ,$$
$$\tilde{\varphi}_t - f\tilde{\theta} + ikc_0^2\tilde{\mu} = 0 ,$$
$$\tilde{\theta}_t + f(1 - Q)\tilde{\varphi} + \beta c_0^2 Q\tilde{\mu} = 0 . \quad (2.37)$$

One can verify that the compatibility condition of (2.37), which is obtained after the substitution $\tilde{\varphi} = \exp(\lambda t)\varphi$, gives the dispersion relation that corresponds to (2.33) with $\lambda = -i\sigma$. The relationship between the vector components $\Phi^T = \{\tilde{\mu}, \tilde{\varphi}, \tilde{\theta}\}$ satisfying an equation of the same type as (2.29)

$$\Phi_t + L(ik)\Phi = 0 \tag{2.38}$$

is determined by equalities

$$\tilde{\varphi}_i = \frac{f\sigma_i + k\beta c_0^2}{\beta\sigma_i + kf}\tilde{\mu}_i \equiv a_i\tilde{\mu}_i ,$$

$$\tilde{\theta}_i = \frac{\sigma_i^2 - k^2 c_0^2}{i(\beta\sigma_i + kf)}\tilde{\mu}_i \equiv b_i\tilde{\mu}_i . \tag{2.39}$$

One can verify the correctness of the formulas once more by putting (2.39) into (2.37) and getting the dispersion relation. The general structure is standard

$$\Phi_i = \begin{pmatrix} 1 \\ a_i \\ b_i \end{pmatrix} \tilde{\mu}_i \tag{2.40}$$

where a_i, b_i are obtained by the substitution of σ_i from (2.34,35) into (2.39) and where $\tilde{\mu}_i$ are integration constants which are determined by the initial conditions.

Thus the hydrodynamical variables that are the vector components Φ_i for the Rossby or Poincaré waves are related both through their amplitude and in their phase which allows one to identify the contributions of each wave in mixed excitation. Commonly $\Phi = \sum_{i=1}^{3} \Phi_i = \sum_{i=1}^{3} P^{(i)}\Phi$.

We build up a general form for the projection operators to the subspaces of (2.40) with definite values a_i and b_i. By calculation we state that if

$$P^{(i)} = \begin{pmatrix} \alpha_i & \beta_i & \gamma_i \\ a_i\alpha_i & a_i\beta_i & a_i\gamma_i \\ b_i\alpha_i & b_i\beta_i & b_i\gamma_i \end{pmatrix} , \tag{2.41}$$

then $P^{(i)}(A, B, C)^T = \mu_i(1, a_i, b_i)^T$; $\mu = \alpha_i A + \beta_i B + \gamma_i C$. The condition $[P^{(i)}]^2 = P^{(i)}$ means that

$$\alpha_i + a_i\beta_i + b_i\gamma_i = 1 .$$

Thus β_i, γ_i may be considered as parameters which are determined by the orthogonality conditions ($i \neq k$)

$$P^{(i)}P^{(k)} = 0 . \tag{2.42}$$

Namely, if $\Delta \equiv [b \times a]$: $\Delta_1 = b_2 a_3 - b_3 a_2$, $\Delta_2 = b_3 a_1 - b_1 a_3$, $\Delta_3 = b_1 a_2 - b_2 a_1$, $\Xi = \Delta_1 + \Delta_2 + \Delta_3$, $\alpha_i = \Delta_i/\Xi$, $\beta_1 = (b_3 - b_2)/\Xi$, $\beta_2 = (b_1 - b_3)/\Xi$, $\beta_3 = (b_2 - b_1)/\Xi$, $\gamma_1 = (a_2 - a_3)/\Xi$, $\gamma_2 = (a_3 - a_1)/\Xi$, $\gamma_3 = (a_1 - a_2)/\Xi$, then the orthogonality conditions (2.42) are fulfilled. The projection operators may be represented in the form of (2.40) with the condition (2.42) for problems of any dimensionality: $P_{ik}^{(m)} = \alpha_{mi}\pi_{mk}$, where $\pi^T\alpha = I$ [2.16].

One can verify that the projection of (2.38) in coordinate representation gives the system of independent equations

$$W_t^{(\alpha)} + i\sigma_\alpha(-i\frac{\partial}{\partial x})W^{(\alpha)} = 0, \qquad (2.43)$$

where $W^{(\alpha)} = P^{(\alpha)}\Phi$ are the scalar functions (all the components of the equation are equivalent) and

$$\mu = \sum_\alpha W^{(\alpha)}, \quad \varphi = \sum_\alpha a_\alpha W^{(\alpha)}, \quad \theta = \sum_\alpha b_\alpha W^{(\alpha)}. \qquad (2.44)$$

2.3 Interaction of Surface Poincaré and Rossby Waves

The basis of the description of wave interaction is the nonlinear equation (2.32). Using the relations $\phi = r\varphi + s\mu$, $\nu = r\mu - c_0^2 s\varphi$ where $r = \beta c_0^2/(f^2 - \beta^2 c_0^2)$, $s = fr/\beta$, and reverting to the mode indices θ, φ, μ let us write the series (2.36) and project equations (2.32) to the basis subspaces by scalar multiplication of each equation by the functions Y_n:

$$\mu_t^n + \varphi_x^n - \beta\theta^n = A^n \equiv (Y_n, A) = \int_0^L e^{-\beta y} Y_n A \, dy$$

$$\varphi_t^n + c_0^2 \mu_x^n - f\theta^n = B^n$$

$$\theta_t^n + f(1-Q)\varphi^n + c_0^2 \beta Q \mu^n = C^n. \qquad (2.45)$$

The expressions for A^n, N^n, C^n are given in Appendix 2. In order to write the equation of interaction we use the projection operators (2.40) on (2.45). The left side of the equation will have the form of (2.43). In the right side there will be a combination of the right sides of (2.32):

$$W_{nt}^{(\nu)} + i\sigma_\nu(-i\partial/\partial x)W_n^{(\nu)} = \alpha_\nu A^n + \beta_\nu B^n + \gamma_\nu C^n. \qquad (2.46)$$

For illustration we shall consider the case of weakly nonlinear long waves ($k \to 0$). Expanding $\sigma_i(k)$ in a Taylor series in k up to third order we have

$$\sigma_1(k) = \frac{\beta f k}{l_n^2 + f^2/c_0^2}\left[1 - \frac{k^2}{l_n^2 + f^2/c_0^2}\right],$$

$$\sigma_{2,3}(k) = \pm c_0\left(\sqrt{l_n^2 + f^2/c_0^2 + \beta^2/4} + \frac{k^2}{\sqrt{l_n^2 + f^2/c_0^2 + \beta^2/4}}\right).$$

In the projection operators we shall leave only the contributions with $k = 0$ because nonlinear terms are small due to the relative smallness of the amplitude. The explicit form of the operators is given in Appendix 2. Putting them into

(2.46) we shall get the approximate interaction equations that are valid for long waves and small amplitudes,

$$W^{(1)}_{nt} + i\sigma_1(-i\partial/\partial x)W^{(1)}_n = (fA^n - \beta B^n)/(f - \beta a_1),$$
$$W^{(2,3)}_{nt} + i\sigma_{2,3}(-i\partial/\partial x)W^{(2,3)}_n = \frac{B^n - a_1 A^n}{2(f/\beta - a_1)} \pm \frac{C^n}{2b_2}. \quad (2.47)$$

Let the only excited transverse mode be n and let it contain only one Rossby and one of the Poincaré waves $W^{(3)}$. Denote $W^{(1)}_n \equiv R$, $W^{(2)}_n \equiv \Pi$, $A^n \equiv A$, $c_R \equiv \beta f/(l_n^2 + f^2/c_0^2)$, $d_R = c_R/(l_n^2 + f^2/c_0^2)$, $a^{(1)}_n = a$, $b^{(2)}_n = b$, $\sigma_3(0) = \sigma_n = c_0 / \sqrt{l_n^2 + f^2/c_0^2 + \beta^2/4}$, then

$$\sum_\alpha W^{(\alpha)}_n = R + \Pi, \quad \sum_\alpha a^{(\alpha)}_n W^{(\alpha)}_n = aR + f\Pi/\beta,$$
$$\sum_\alpha b^{(\alpha)}_n W^{(\alpha)}_n = b\Pi. \quad (2.48)$$

In the right-hand sides of the equations we leave only nonlinear terms without derivatives and contributions of the Poincaré waves which are linear in β and are contained in C^n. By (A2.6) denoting

$$c_\Pi = [(Y_n, y(Y_n + Y'_n r))f + \beta s(Y_n, yY')]/[4H_0(f/\beta - a)b],$$

we get (Appendix 2)

$$\Pi_t - i\sigma_\Pi \Pi + c_\Pi \Pi_x + c_0^2 \Pi_{xx}/2\sigma_\Pi$$
$$= -\frac{\beta\pi}{2H_0(f - \beta a)}\left[(n, n, n')b^2\Pi - aA\right], \quad (2.49)$$
$$R_t - c_R R_x - d_R R_{xxx}$$
$$= -\frac{\beta\Pi}{H_0(f - \beta a)}\left[\frac{f}{\beta}A - sb(n, n', n)(R + \Pi) + b^2(n, n, n')\Pi\right]$$
$$A = \left[(1 + \beta r)(n, n, n') - (l^2 + \beta^2/4)(n, n, n)\right](aR - f\Pi/\beta)$$
$$+ \left[s\beta(n, n, n') - s(l^2 + \beta^2/4)(n, n, n)\right](R + \Pi). \quad (2.50)$$

Let us derive the equations for the intermode interaction of Rossby waves. We assume that the Poincaré waves are not excited and denote $W^{(1)}_n \equiv R_n$, $W^{(2,3)}_n = 0$, $a^{(1)}_p = a_p$. In this case only the first group of equations (2.47) remains. The expression for the right-hand side is the corollary from (A2.4-5):

$$R_{nt} - c_R R_{nx} - d_R R_{xxx}$$
$$= -\frac{f}{H_0(f - \beta a)}\sum_{pq}\Big(\{(n, p, q) + r[(n, p', q) + (n, p, q')]$$
$$+ r^2(n, p', q')\}a_p a_q + s[(n, p, q') + r(n, p', q')]a_q + s^2\Big)R_p R_{qx}$$
$$= \sum_{p,q} N^n_{p,q} R_p R_{qx}. \quad (2.51)$$

The system (2.51) is the system of coupled Korteweg–de Vries (CKdV) equations. The approximate solutions of the system are demonstrated in Sect.2.5,6. It is interesting to note that the sign of the dispersion coefficient in (2.51) is negative. We may suppose that the interaction of a Rossby wave and a long internal gravity wave is described by the coupled system of KdV equations with opposite signs of the dispersion constants. One of the versions of such a system has a Lax pair and a soliton solution.

The expressions (2.49–50) demonstrate one of the fundamental peculiarities of the application of the model evolution equation method in waveguide theory – the systems are infinite. This results from the intermode interaction. However, investigations show that in the case of a small mode number excitation, the weak nonlinearity, the finite space-time area of interaction, allows one to construct the effective nonlinear wave description [2.23–26]. From this example it is clear that the computation of nonlinear constants is difficult. Their computation is possible by an analytic transformation program such as REDUCE. We add that the model KdV equations for planetary waves were introduced in [2.27], see also [2.28].

2.4 Long Atmospheric Internal Waves

The first attempts at theoretical investigation of nonlinear internal waves were made by *Long* [2.29] and *Benjamin* [2.30]. They considered an incompressible fluid with the equation of state given in (2.9). The results of their work were the derivation of the stationary KdV equation for an isolated mode and the analysis of its simplest solutions. Analogous results for an atmospheric compressible fluid with the state equation (2.7) were obtained by *Pelinovsky* and *Romanova* in [2.31]. The development of this theory for nonstationary nonlinear wave mode evolution in fluids is given in [2.32–34]; nonstationary nonlinear atmospheric waves are considered in [2.35,36]. The multi-mode problem without mode interaction but with turbulent dissipation was solved in [2.37]. The problem of internal wave modes interaction was studied in [2.24,38,39]. One could also note the numerical experiments on generation and propagation of long internal waves in the hydrostatic approximation [2.12,40,41]. In this approach the vertical acceleration term is neglected, which is equivalent to the approximation of nondispersive mode evolution. The natural background state and the inclusion of viscosity terms allow one to make clear the important features in nonlinear wave behavior. We shall use these features in the statement of the problem and in choosing the system of small parameters. First of all, we can determine the upper boundary condition: the tendency of the dynamical variables to approach a constant value occurs in the region of the upper bound. In addition there is the intermode capture effect – the appearance of a relatively stable contribution which is supported by the intermode interaction [2.25,41]. The exposition below is based on the results of [2.25,26,39,42–46] with addition of effects of earth rotation.

2.4.1 Internal Waves in an Atmospheric Waveguide

The statement of the problem of generation and propagation of internal waves in the atmosphere includes the system of equations (2.7,8). Restricting ourselves to internal waves, we can simplify the formulas without reference to the cumbersome procedure using projection operators of the fifth order. Neglecting the dependence of the Coriolis parameter on the latitude we exclude the Rossby waves. The choice of the hydrostatic approximation as the zeroth approximation excludes the acoustic contribution [2.4]. Thus, in the third equation of the vector subsystem (2.20) we bring w_t to the right-hand side, including the vertical acceleration in b_3. After operations leading to (2.25,26) we get this system without terms describing the horizontal dispersion [derivatives of the fourth order in $(x - t)$]. Hereafter the spherical coordinates (r, ϑ, φ) are introduced. Restricting ourselves to the first term of the ϑ-expansion in the Coriolis acceleration near the point $\vartheta = \vartheta_0$ and excluding δ we get (A1.3)

$$\left[\frac{\gamma}{\gamma-1}\left(\frac{\partial}{\partial r} - \frac{\partial^2}{\partial r^2} - \frac{2(\gamma-1)}{\gamma r}\right)\frac{\partial^2}{\partial t^2} - \Delta_\perp + \frac{1}{\gamma-1}\frac{\partial^2}{\partial t^2}\left(\frac{\partial^2}{\partial t^2} - \gamma \Delta_\perp\right)\right.$$
$$\left. + \frac{2(2-\gamma)}{\gamma-1}\Omega\frac{\partial^2}{\partial \varphi \partial t} + \frac{4\Omega}{r}\frac{\partial}{\partial \varphi}\int_0^t -\frac{\gamma f^2}{\gamma-1}\left(\frac{\partial^2}{\partial r^2} - \frac{\partial}{\partial r}\right)\right]T$$
$$= \int dt^5 dr [C^2 F + ACG - 2CAG] \tag{2.52}$$

where $\Delta = \Delta_r + (1/r^2)\Delta_{\vartheta\varphi}$. When $\Omega = 0$ the left side in (2.52) is explicit. Let us assume that we are investigating the long internal wave propagating along the planet surface. In a local coordinate system with the zero point at $r = a$, $\vartheta = \vartheta_0$, $\varphi = 0$ we set the x axis to be along the direction of increasing latitude ϑ and the y axis in the orthogonal direction, along the parallel. Such a geometry is convenient for describing thermospheric internal wave generation in the auroral electrojet [2.40,41,47,48]. The solution of the problem is given as an example below.

The mathematical description of the evolution of long weakly nonlinear internal waves will be developed by using small dispersion parameters. The critical parameter of the horizontal mode dispersion is the relation of the height to the wavelength along the x axis propagation: $\beta = H/\lambda_x$. Let also $\lambda_y \gg \lambda_x$, i.e., there exists one more small dispersion parameter $\nu = \lambda_x/\lambda_y$. We suppose further that the nonlinearity scale σ is determined by the ratio of the temperature variation T to the background temperature $\bar{T}$ meaning $\sigma = \max T/\bar{T}$. The speed of dissipation of the wave disturbance depends on the magnitude of the coordinate derivatives and on the dissipation constants of the medium. It has been shown for long waves that for modes with small numbers in the atmosphere [2.25,49] and the ocean [2.15,37] the dissipation may be taken into account by perturbation theory (Sect. 6.2). Depending on the dimensions of the wave and space-time propagation intervals, spherical curvature and rotation of the planet might have to be taken into account [2.15,47,50]. The natural small parameters are now a/H where a is the planet radius and the dimensionless angular frequency modulus is

$\Omega\sqrt{H/g}$. It is clear that the contribution of these planetary effects will be large if the horizontal scales of disturbances are comparable with a and the lifetimes of characteristic processes are at least a few hours. These waves are classified as large-scale internal waves and they play an important role in the physics of the atmosphere-ionosphere. The propagation of these waves cause strong variations in the state of the ionosphere during magnetospheric substorms [2.51] and influences the mean parameters of the atmosphere and its energetics [2.52,53]. The long action of the mid-scale internal waves results in similar effects [2.54,55].

In order to develop a perturbation scheme in (2.52) we make the following substitution $t \to t/\beta$, $y \to y/\nu\beta$, $x \to x/\beta$, $z \to z$. This allows one to estimate the functions and their derivatives [2.8,23]. The relations between dynamical variables in the zero order of the small parameters are

$$u \to \sigma u, \; v \to \beta\nu\sigma v, \; w \to \sigma\beta v, \; T \to \sigma T, \; \delta \to \sigma\delta \, .$$

In fact this change of variables formalizes the assumed initial conditions and the hypothesis of the solution behavior. The choice of small parameters and the relations between them in the nonlinear theory is especially important because of the existence of nonlinear resonance – the effects of accumulation of small disturbances. The appearance of resonances usually occurs in the zero and first approximations, that is, linearization relative to the background. Nonlinear resonances occur in the increase of first order corrections with time while linear ones occur due to the so-called instability. The gist of the method lies in the account of all the resonances of the model system in the approximation of the first order perturbation theory.

If all the terms from (2.52) containing small parameters were put in the right-hand side, then on the left-hand side there would only be contributions of the operator containing derivatives of the first and second order in the coordinates and in time:

$$LT = \Delta_r \left(\frac{\gamma}{\gamma - 1} T_{tt} + T \right) + \frac{1}{r^2} \Delta_{\vartheta\varphi} T$$

$$- \left(\frac{\partial^2}{\partial t^2} + \frac{\partial}{\partial r} \right) \left(T_r + \frac{2}{r} T \right) - \frac{1}{\gamma - 1} T_{ttr} = F \, . \tag{2.53}$$

As an illustration let us give the expression for the nonlinear term in the heat equation (2.21):

$$d \sim -\beta \left[\frac{(uT_x + \nu^2 \beta v T_y + w T_z)}{(\gamma - 1)} + T(u_x + \nu^2 \beta v_y + w_z) \right] \, . \tag{2.54}$$

In the zeroth approximation the relations between the dynamical variables may be obtained directly from (2.19–21) and are

$$\delta = \int_a^r T \, dr - T + C_1(\vartheta, \varphi, t) \, , \tag{2.55}$$

$$w = -\frac{\gamma}{\gamma - 1}T_t + \int_a^r T_t dr + C_{1t}(\vartheta, \varphi, t) \qquad (2.56)$$

$$u = -\frac{1}{r}\frac{\partial}{\partial \vartheta} \int_0^t \left[\int_0^r T dr + C_1 \right] dt + C_2(\vartheta, \varphi, r) \qquad (2.57)$$

$$v = -\frac{1}{r \sin \vartheta}\frac{\partial}{\partial \varphi} \left[\int_0^t T dr + C_1 \right] dt + C_3(\vartheta, \varphi, r) . \qquad (2.58)$$

The constant C_1 appears after integrating the vertical component equation (2.20) in the linear approximation, and is determined by the boundary conditions. The functions $C_{2,3}$ coincide with the initial values of the horizontal velocity projections. The substitution (2.55–58) into (2.19–21) leads to the condition for C_k, which results from the invariant relation in (2.19-21) that defines the internal wave. Relations (2.55-58) close the nonlinear expressions such as (2.54) in the first order approximation.

Let us continue to discuss the statement of the problem. Equation (2.53) describes long internal waves and contains only derivatives of second order for any independent variables in the left-hand side, which defines the statement of the problem. Therefore, it is necessary to give two initial conditions and two boundary conditions for every variable. We allow that at infinite x the solutions should decrease. Let us suppose that the initial conditions or sources working during a finite time are concentrated around the colatitude ϑ_0 and occupy a space of dimension $\sim \lambda_x$. According to the meaning of the coordinate y being proportional to the longitude φ we shall set the periodical condition $T(\varphi + 2\pi) = T(\varphi)$. As was mentioned above, the choice of the boundary condition for the vertical coordinate $r = a + z$ depends on the atmospheric waveguide to be considered. On the solid horizontal boundary we put $w = 0$. Then $C_1 = \gamma T|_{r=a}/(\gamma - 1)$. At the lowest waveguide situated close to the earth's surface usually the upper condition $w = 0$ ($T = 0$) is chosen [2.12,31]. (The upper boundary conditions are discussed in Sect. 5.4 more throughly.) For the thermospheric waveguide, as shown in Sect. 5.4, the upper boundary condition of the third kind is

$$\gamma w_z \big|_{r=a+h} = w . \qquad (2.59)$$

In this chapter, we will show the results of calculations for the boundary conditions $T_z|_{r=a+h} = 0$, $T|_{r=a+h} = 0$, where h denotes the vertical height of the waveguide. The condition on the upper boundary of the atmosphere models the smallness of the heat flux to the outer space around the earth due to the increase in the molecular mean free path. Such a method is used for the calculation of the nonlinear constants, vertical wave vector projection, and the propagation velocity. This calculation can be compared with the results of calculations with the boundary condition (2.59).

Let us go to the determination of the basis functions of the waveguide modes. Separating the variables in the homogeneous equation (2.53) $LT = 0$: $T = R\theta$, $R = r^{-1} \exp(r/2) f(r)$ we get

$$f'' = \left(\frac{1}{4} + \frac{\gamma-1}{\gamma+1}\frac{1}{r} + \frac{\lambda}{r^2}\right) f, \qquad (2.60)$$

$$\lambda \theta_{tt} + \frac{\gamma}{\gamma - 1}(\Delta_{\varphi\vartheta} + 2)\theta = 0. \qquad (2.61)$$

Let $h \ll a$. This condition is obviously valid on the earth. The linear operator and basis functions can be simplified. To a very good approximation (2.60) gives

$$f'' = \left(\frac{1}{4} + \frac{\lambda}{a^2}\right)f. \qquad (2.62)$$

The boundary conditions (2.59), or any other homogeneous conditions, determine the spectrum $\lambda_n = a^2/c_n^2$. The deviation from the true values, due to spherical curvature, in the most unfavorable case is less than half a percent. To the same degree of accuracy, in a belt of 3000 km width around ϑ_0 in the middle latitudes, one can change (2.61) by

$$\theta_{tt}^n - c_n^2 \left(\frac{\partial^2}{\partial x^2} + \frac{\partial^2}{\partial y^2} + \frac{\cot \vartheta_0}{a}\frac{\partial}{\partial x}\right)\theta^n = 0,$$

where $x = a(\vartheta - \vartheta_0)$, $y = \varphi a \sin \vartheta_0$, n is an eigenvalue. The squared propagation velocity in the zeroth approximation is equal to

$$c_n^2 = \frac{4\gamma}{(\gamma-1)(1+4k_n^2)},$$

where k_n is the vertical component of the wave vector which can be calculated from the boundary conditions.

Let us turn back to (2.53). The function F contains both nonlinear terms N, dispersion, dissipation terms D and Coriolis terms K, $F = N + D + K$, where for K see (A1.3) and the main part of D in small parameters is

$$D = \frac{\beta^2}{\gamma-1}T_{tttt} - \frac{\beta^2 \gamma}{\gamma-1}\Delta_\vartheta T_{tt} + \nu_0[\nu u_{zz} + (\nu - \nu_z)u_z]_{zx}$$
$$+ \frac{\xi_0}{\gamma-1}\left\{[\xi(T_z - T)]_{zz} + (\xi - \xi_z)T_z\right\}_{tz}. \qquad (2.63)$$

Expression (2.63) is obtained if in the dissipative terms of (2.5–19) only the contributions of the first order are retained. Showing by the arrow the transition to dimensionless variables, we write in the designations of (2.19–21)

$$b_1 \sim \frac{1}{\varrho}\frac{\partial}{\partial z}\eta\frac{\partial u}{\partial z} \to \frac{1}{\varrho_0}\frac{\partial}{\partial z}\varrho_0 \nu'\frac{\partial u}{\partial z},$$

$$d \sim \frac{1}{\mu\varrho_0}\frac{\partial}{\partial z}\kappa\frac{\partial T}{\partial z} \to \frac{1}{\varrho_0}\frac{\partial}{\partial z}\varrho_0\xi'\frac{\partial T}{\partial z},$$

where

$$\nu'(z) = \frac{\eta(z)}{\beta H \sqrt{gH} \varrho_0(z)} = \nu_0 \nu ,$$

$$\xi'(z) = \frac{\kappa(z)\mu}{\beta c_v H \sqrt{gH} \varrho_0(z)} = \xi_0 \xi , \tag{2.64}$$

moreover, $\max\{\xi, \nu\} = 1, \nu_0; \xi_0 \ll 1$ in the height interval $[0, h]$.

The nonlinear terms expressed through T by means of the linear relations (2.55–58) allow one to write the explicit form of (2.53) to first order of the small parameters $\sigma, \beta, \nu_0, \xi_0, h/a, \Omega/\beta$:

$$\begin{aligned}
N = & [T(T - T_z)]_{ztt} - \left[\int_0^t\!\!\int_0^z T_{xx}dz\,dt + \int_0^t\!\!\int_0^z T_x dz\,dt \right.\\
& \left. - \left(\int_0^z T dz - \frac{\gamma-1}{\gamma}T \right)_t \int_0^t T_x dt + T \left(\int_0^z T dz - T \right)_x \right]_{xz}\\
& + \left\{ \frac{1}{\gamma-1}\left[T_z \left(\int_0^z T dz - \frac{\gamma}{\gamma-1}T \right)_t - \int_0^t\!\!\int_0^z T_x dz\,dt\,T_x \right] \right.\\
& \left. + T \left(T_t - \frac{\gamma}{\gamma-1}T_{tz} - \int_0^t\!\!\int_0^z T_{xx} dz\,dt \right) \right\}_{tz} + [T(T - T_z)]_{xx}\\
& + \left\{ (T - T_z)\left(\int_0^z T dz - \frac{\gamma}{\gamma-1}T \right) - \int_0^t\!\!\int_0^z T_x dz\,dt \left(\int_0^z T dz - T \right) \right\}_x\\
& - \frac{1}{\gamma-1}\left[T_z \left(\int_0^z T dz - \frac{\gamma}{\gamma-1}T \right)_t - T_x \int_0^t\!\!\int_0^z T_x dz\,dt \right]\\
& \left. - T \left[T_t - \frac{\gamma}{\gamma-1}T_{zt} - \int_0^t\!\!\int_0^z T_{xx} dz\,dt \right] \right\}_{zzt}.
\end{aligned} \tag{2.65}$$

Details of the derivation of (2.53) at $\Omega = 0$ can be found in [2.47]. A discussion of the next higher order of perturbation theory is given in Sect. 2.5.

The solution of (2.53) is represented by a Fourier expansion with the basis functions $Z_n = E_n^{-1/2} e^{z/2} \sin k_n z$, where Z_n are normalized functions proportional to $R(a + z) = e^{z/2} f(a + z)$ that satisfies the boundary conditions $Z_n(0) = R_n(a) = 0, \gamma Z_n = Z_n(h) : E_n = h/2 + 1/(4k_n^2 + 1)$. Thus

$$T = \sum_{n=1}^{\infty} \theta^n(x,y,t) Z_n(z) . \tag{2.66}$$

Putting (2.66) into (2.53) and projecting the result over the subspaces of the basis functions Z^n in analogy with (2.45) we get the interacting mode system

$$\theta_{tt}^n - c_n^2 \left(\Delta_\perp + \frac{\cot \vartheta_0}{a} \frac{\partial}{\partial x} \right) \theta^n$$

$$= -\frac{2(\gamma-1)}{\gamma(2h+\mu_n)^{1/2}} \int_0^h (N + D + K) e^{-z/2} \sin k_n z\, dz , \tag{2.67}$$

where $\mu_n = 4/(4k_n^2 + 1) = c_n^2(\gamma - 1)/\gamma$. Equation (2.67) contains the second order derivatives in t, x, y on the left-hand side and the fourth order ones on the right-hand side. Moreover, the functions θ^n, after the substitution of (2.66) into the nonlinear part of (2.65) enter into the integrands and various derivatives so that the right-hand side looks rather complex. Usually [2.33,43], with the aim of further simplification of the equations, new variables $\xi = x - c_n t$, $\tau = \sigma t$ are introduced and, for monodirectional waves, contributions of σ^2 order are neglected. In perturbation theory this is equivalent to the existence of a closed system to first order in t and x which gives, in the zeroth approximation, $\theta_t^n = \pm c_n \theta_x^n$. In two-dimensional ($xy$) mode systems, such a confirmation is not generally valid. Equation (2.67) is the second order in t and, due to θ_{yy}^n, the waves do not split so easily. Here the wave front can assume a curvature of the σ-th order.

Therefore we return to the initial equations of the first order and use projections to the mode basis. For simplicity we neglect rotation of the earth, dissipation, and curvature. These can be taken into account in the linear operator terms that are given in (2.53). First we develop the mode expansion and project the solution over the subspaces of the left- and right-directed waves by the method described in [2.7,16]. In the case of the directed wave one can use the slow time formalism making the substitutions in the small parts of (2.67) and changing all the derivatives with respect to x to ones to t. Let us do this for all the equations except the y-projection of (2.20). In correspondence with the arrangement of z-derivatives in the y-projection, the expansion for v_z is $\partial v / \partial z \equiv v_z = \sum_n f^n(x, y, t) Z^n(z)$. We get

$$\theta_t^n + c_n \theta_x^n - \sigma \sum_{m,k} \frac{\Phi_{km}^n \theta^k \theta_t^m}{c_m} - \frac{\beta^2 M^n \theta_{ttt}^n}{2c_n^3} + \frac{\nu^2 c_n^2 f_y^n}{2} = 0 . \tag{2.68}$$

The equation for the projection of (2.20) gives

$$f_t^n + \theta_y^n = 0 . \tag{2.69}$$

Differentiating (2.68) with respect to t and substituting (2.69) into the result we get

$$\left[\theta_t^n + c_n \theta_x^n - \sigma \sum_{m,k} \frac{\Phi_{mk}^n \theta^k \theta_t^m}{c_m} - \frac{\beta^2 M^n \theta_{ttt}^n}{2c_n^3} \right]_t - \frac{\nu^2 c_n^2 \theta_{yy}^2}{2} = 0 . \tag{2.70}$$

If one neglects the mode interaction [i.e., sets in (2.70) $\Phi_{km}^n = \Phi^n \delta_{nk} \delta_{nm}$] the Kadomtsev–Petviashvili equation [2.56] will be obtained. When this equation is compared with the usual one for internal waves [2.24,56,57], one sees that the x and t variables are interchanged [2.58]. Naturally the t–derivatives in (2.68) in the nonlinear and dispersion terms may be changed to the x–derivatives with the same degree of accuracy but the change $\partial/\partial t$ in a linear operator is not possible. Therefore the theory is valid only over the solution set for which

$u_y - v_x = \text{rot}_z \boldsymbol{v} = 0$ with an accuracy of $O(\sigma)$ [2.56,57]. In this case $f_t^n \simeq -c_n f_x^n$ and the t–derivative of θ^n can be changed to θ_x^n. The condition $\text{rot}_z \boldsymbol{v} = 0$ is approximately true if the initial condition also satisfies it. Note that $\text{rot}_z \boldsymbol{v}$ is the integral of motion ($\Omega = 0$) in the linear theory.

Equation (2.70) is the generalized KP equation for the multi-mode situation and the two-dimensional generalization of the KdV system (CKdV). Methods of CKdV solution are discussed in Sect. 2.5. Thus, if the mode interaction can be neglected the internal wave evolution is described by the KP equation, which allows one to find the two-dimensional instability and get new stationary solutions. The three-dimensional instability can be studied by the KP system.

In the large-scale internal wave global propagation problem there is exact periodicity at $y = \varphi a \sin \vartheta_0$. This condition could be satisfied if one uses the results of *Dubrovin* where the solutions are expressed in terms of Riemann theta-functions of genus 2,3 [2.59,60] and *Bobenko* where the automorphic functions are effectively used [2.61] [see also Refs. 2.62 and for an alternative approach Refs. 2.63,64]. A program for calculation of θ^n with the given parameters as a function of coordinates has been elaborated. An analogous program is described in [2.65]. Figure 2.1 gives a representation of the solution properties. In part (a) the time evolution of the KP solution at $x, y = 0$ is presented. In parts (b,c) the dependence on x and y is given. The values of the parameters and units are for large-scale thermospheric internal waves that are generated by the auroral electrojet [2.41,45,47]. We list here the explicit expressions for the dispersion and nonlinear constants:

$$M^n = \frac{\mu_n c_n}{2}(1 - c_n^2/\gamma)$$

$$\Phi^n = \frac{c_n k_n}{\gamma}\left[k_n^2 \mu_n^2 \lambda_n \eta_n \frac{5 - 6.5\gamma - 6(3\gamma + 2)k_n^2}{h - (\gamma - 1)(3\gamma - 2)/2}\right],$$

where

$$\mu_n = 4/(4k_n^2 + 1), \quad \eta_n = (-1)^n e^{h/2} - 1, \quad \lambda_n = 4/(36k_n^2 + 1).$$

The nonlinear constants are calculated for those Z^n that satisfy the zero conditions at the boundaries. The values of the first modes of the parameters are given in Table 2.1. The choice of dimensionless variables sets $\sigma = \beta = 1$.

Table 2.1. Nonlinear constants corresponding to zero boundary conditions for $n = 1 \ldots 4$.

n	1	2	3	4
k_n	0.45	0.90	1.34	1.79
Φ^n	1.12	-0.97	0.40	-0.47
d^n	0.00	0.00	0.02	0.08
c_n	0.86	0.56	0.40	0.31
M_n	0.48	0.21	0.09	0.04

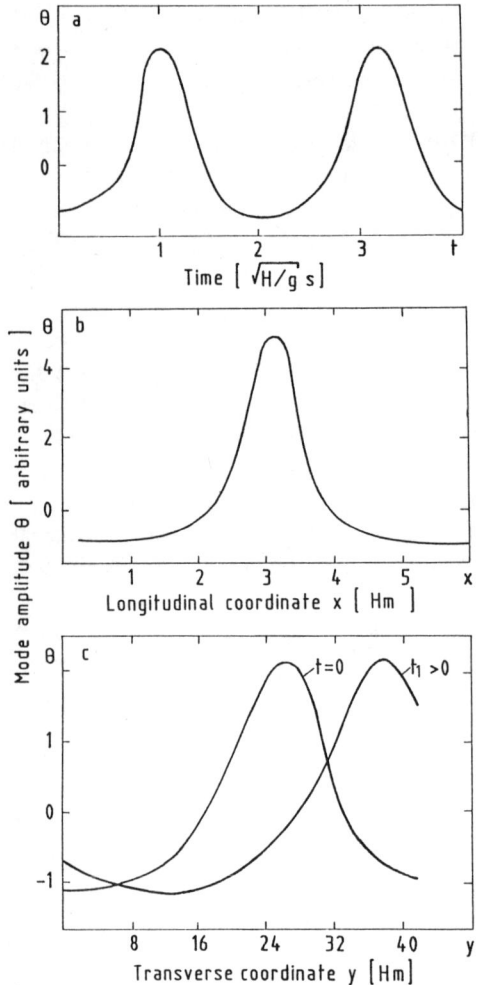

Fig. 2.1. Finite-gap genus 2 (Riemann θ-function) solution of the KP equation. (a) Time dependence $\theta(0,0,t)$. (b) Longitudinal coordinate dependence $\theta(x,0,0)$. (c) Transverse coordinate dependence $\theta(0,y,0)$ and $\theta(0,y,t_1)$

Effects of Dissipation. The influence of viscosity and thermoconductivity can cause mode quenching as well as mode interaction, and increases with the mode number [2.25,45]. Taking into account the dissipation in the two-dimensional case leads to the CKdV with the damping terms

$$\theta_t^n + c_n \theta_x^n + \sigma \sum_{mk} \Phi_{mk}^n \theta^m \theta_x^k + \beta^2 M^n \theta_{xxx}^n + \tilde{\nu} \sum_l d_l^n \theta^l = 0 \tag{2.71}$$

$$n, l, m = \pm 1, \pm 2, \ldots, \quad c_{|n|} = -c_{|n|},$$

where $\tilde{\nu}$ denotes the maximum of the small parameters ν_0, ξ_0 in (2.64). The numerical values of the constants d_l^n are calculated in [2.25]. In Table 2.1 the diagonal elements of $d^n \equiv d_n^n$ are given.

The simplest method of accounting for the nonlinearity, dispersion and dissipation leads to a KdV equation with damping for each mode [2.66,67]. Setting $\sigma = \beta = \tilde{\nu} = 1$ and dropping the indices let us write

$$\theta_t + c\theta_x + \Phi\theta\theta_x + M\theta_{xxx} + d\theta = 0 \ . \tag{2.72}$$

The values of the constants for a thermospheric waveguide are included in Table 2.1. By means of (2.72) it is possible to analyze the role of the hydrostatic approximation $w_t = 0 \sim M^n = 0$ in large-scale internal waves. Transforming $x \to x - ct$, $t \to \tau$, $\theta = \tilde{\theta}\exp(-d\tau)$ we then have

$$\tilde{\theta}_\tau + \Phi\exp(-d\tau)\tilde{\theta}\tilde{\theta}_x + M\tilde{\theta}_{xxx} = 0 \ . \tag{2.73}$$

Let us first suppose that $M = 0$. Then the waveguide mode is described by the quasilinear equation

$$\tilde{\theta}_\tau + \Phi\exp(-d\tau)\tilde{\theta}\tilde{\theta}_x = 0 \ . \tag{2.74}$$

Equation (2.74) may be solved by the method of characteristics. The solution demonstrates that nonlinearity leads to wave overturning despite the influence of dissipation [$d\theta$ term in (2.72)]. The solution, at sufficiently long time, will not have any physical sense. The correct solution can be found as a piecewise regular function, as is the case in shock wave theory [2.68]. For the piecewise smooth generalized solutions to be the limiting case when the dispersion $M \neq 0$, one needs to take into account

$$\lim_{M \to 0} u(x, t, M) = \tilde{\theta} \ ,$$

where u is the solution of (2.73).

We then use the heuristic reasons for the determination of the class of the solution of (2.73). This equation has a structure analogous to the KdV which, for smooth initial conditions, has smooth solutions. The uniqueness of (2.73) is restricted by the fact that the coefficient before $\tilde{\theta}\tilde{\theta}_x$ decreases with time. Since the phenomenon of the cutoff is caused by nonlinearity, this discrimination should not spoil the smoothness of the solution. This can be supported by the form of the approximate solutions which are listed in [2.66]. For the solution of (2.73) one can derive the conservation laws

$$\frac{dU}{d\tau} = \frac{d}{d\tau}\int_{-\infty}^{\infty} u(x,\tau)dx = 0 \ ,$$

$$\frac{dE}{d\tau} = \frac{d}{d\tau}\int_{-\infty}^{\infty} u^2(x,\tau)dx = 0 \ .$$

For (2.74), without loss of generality, we can consider the partial solution which has a single discontinuity that propagates with the velocity $V = y(t)$, where $y(t)$ is the coordinate along which the discontinuity occurs. The solution corresponds to the conservation laws

$$\frac{d\tilde{U}}{d\tau} = \frac{d}{d\tau}\int_{-\infty}^{\infty}\tilde{\theta}(x,t)dx = \frac{\Phi}{2}e^{-d\tau}[\tilde{\theta}^2] - V[\tilde{\theta}], \qquad (2.75)$$

$$\frac{d\tilde{E}}{d\tau} = \frac{d}{d\tau}\int_{-\infty}^{\infty}\tilde{\theta}^2(x,t)dx = -V[\theta^2] + \frac{2M}{3}[\theta^3], \qquad (2.76)$$

where $[f] = f(y+0) - f(y-0)$. If now we assume the existence of a limiting transition for the solutions (2.73,74) and, as is the case with

$$\frac{d\tilde{U}}{d\tau} = \lim_{M\to 0}\frac{dU}{d\tau} = 0, \quad \frac{d\tilde{E}}{d\tau} = \lim_{M\to 0}\frac{dE}{d\tau}$$

expressing V through (2.75) and inserting into (2.76), we find that the assumption of the existence of a discontinuity is contradicted $Me^{-d\tau}[\tilde{\theta}]^3/6 = 0$. Thus the hydrostatic approximation [2.40] in nonlinear theory is not valid at times longer than the time of wave inversion.

This analysis of the limiting transition shows that in long nonlinear internal wave propagation in a weakly dissipating medium there is a strong difference with the known case of shock sound waves [2.68] and the dissipation cannot compensate for the increasing steepness in the horizontal profile. This is due to $Hk_x \ll 1$ and the dissipation contribution mainly depends on the vertical wave structure. The compensation for the nonlinear increase in the steepness here depends primarily on the mode dispersion. Discontinuous solutions are completely absent.

In conclusion let us illustrate the properties of (2.72) in the example of an approximate quasisoliton which is obtained under conditions of low dissipation [2.35,66]:

$$\theta(x,\tau) = F\operatorname{sech}^2\left[\sqrt{\frac{c}{4M}}(x - c\tau - \int_0^\tau \tilde{c}(\tau')d\tau')\right]. \qquad (2.77)$$

The amplitude F decreases exponentially with time and is proportional to the quasisoliton velocity so that the latter also decreases with time, $F = 3\tilde{c}/\Phi = (3\Phi)^{-1}c_H\exp(-4d\tau/3M)$, where c_H is the initial soliton speed. As is clear from (2.77), the dissipation leads to the asymmetry of the soliton contributions and the appearance of tails with increasing time [2.66].

We note in addition that for pycnoclyne internal waves in the ocean the model evolution equations have many common features with (2.67,70,72). In a weakly stratified fluid (the approximately homogeneous Brent–Väsälä frequency), for long internal waves the system of coupled KP equations is derived too [2.24]. Only the parameter dependence and the values of the constants are different. In Chap. 5 in the example of the incompressible Boussinesq fluid the effects of the strong vertical inhomogeneity due to the existence of thermoclynic waves are investigated. The explicit formulas for the equation coefficients are given. The analogous situation for atmospheric gas is discussed too.

We have studied the quasi-two-dimensional waves with the spatial scales $\lambda_z \ll \lambda_x \ll \lambda_y$. In [2.69] the equations for the nonlinear waves of another geometry are derived, for example, the case of cylindrical symmetry. In this reference there are also rather complete lists of publications and comments. For the Johnson equation inverse problem see [2.70].

2.5 The Method of Successive Approximations. Solution for the Cauchy Problem of the Coupled Korteweg–de Vries Equations

The equation structure and the choice of initial conditions which determine the nonlinearity, dispersion and dissipation parameters are such that they allow one to use an iteration procedure. At each step of the procedure some model system with small parameters is formed. The solutions of these evolution equations have the property that the coordinate dependence changes increase due to the small terms at a finite time and are not small. These changes are displayed in the increase of the steepness of the wave front and in the decay into the soliton train. The range of scale variation should change in such a way that the major parameters remain small during the entire allowed time interval. In other words the time interval at which the solution is valid within the given frame of accuracy should be in accordance with the iteration number and the value of the small parameter. In this case the procedure is asymptotic in the neighborhood of $t = 0$.

The derivation of the model equation for T (2.53) or the model systems (2.49–51,70) is a step towards the construction of an approximate solution to the general wave disturbance problem. The iterations allow one to get the hierarchy of equations, so that at each step of the solution, of e.g. (2.53), the condition $|T^{(n+1)} - T^{(n)}| < \sigma^n$ at $t \in [0, t_\sigma^{(n)}]$ should be valid, when $t_\sigma^{(1)} \approx \sigma^{-1}, t_\sigma^{(2)} \approx \sigma^{-2}, \ldots$ where σ is the major small parameter and n is the iteration number. The relations (2.55–58) in the zeroth linear approximation can be infinitely refined with the help of the basis equations by the substitution (2.55–58) into nonlinear terms. Thus the iteration result is

$$N + D = \sum_{k=1}^{\infty} \sigma^k \hat{F}_k(T), \tag{2.78}$$

where $\hat{F}_k$ contain nonlinear, dispersion and dissipation operators. It should be noted that since for each mode the coefficient $\theta_t^n \approx \pm c_n \theta_x^n$ holds, the projected expressions $\hat{F}(\sum \theta^n Z^n)$ will be the same for all long nonlinear waves. Such simplicity can be attained by including the projection onto a subspace of given physical branches of the dispersion relation, i.e., the given waves of interest.

The mode equation after doing the projection step of (2.31) is

$$\theta_t^n + c_n \theta_x^n = \sigma N^n(\theta^1, \ldots). \tag{2.79}$$

Let us suppose that the initial conditions for the system (2.79) and the major parameter σ are such that $\max |N^n(x,t)| \leq 1$ holds in the time interval $[0, t_\sigma]$. Then after integration of (2.79) over time we get

$$\theta^n = \varphi^n(x - c_n t) + \sigma \int_0^t N^n(x - c_n(t - \tau), \tau) d\tau . \qquad (2.80)$$

Equation (2.80) can be interpreted in the following way: at time σ^{-1} the contribution of the weak effects can be nonneglible. The simple perturbation theory, as is well known, cannot improve the accuracy of the result $|\theta^n - \varphi^n| \leq \sigma \int_0^t |N^n| d\tau \leq \sigma t$ in calculating the maximum. Let the iteration in equation (2.79) be carried out as in (2.78) and the iterated equation have smooth solutions with the condition $\max |N^{(2)}(x,t)| \leq 1$ at $t \in [0, t_{\sigma^2}]$. Then the estimated results derived from perturbation theory, made on the basis of a model equation of the first iteration (e.g., KdV) allow the improved solution to differ from the exact one for $\sigma^2 t/2$. The possible increase of the terms on the right-hand side is determined only by the increase in the number of the terms which is dependent on the nature of the specific problem. Therefore at a given time $t \leq 1/2\sigma$ the solution of

$$\theta_t^n + c_n \theta_x^n + \sigma N^{n(1)}(\theta) = \sigma^2 N^{n(2)} + \ldots$$

differs from the solution of the equation with a zero right-hand side by a value that does not exceed σ. In Sect. 2.6 we consider an example of such a second iteration. Thus the iteration scheme using an interpretation of the time dependent behavior as a function of the initial conditions and the solution, together with the mode representation and the polarization projection allow one to investigate complex nonlinear problems.

A necessary property of the equations generated at each iteration step is the boundedness of the solutions as well as of the allowed derivatives at the given time. That is, at each iteration step the entire set of singularities must be attributable to dissipation or dispersion.

Let us demonstrate this on an example of a disturbed KdV. We denote by ψ the solution of the KdV that is taken at $\varPhi = M = 1$, and θ is a solution of the exact equation

$$\theta_t + c\theta_x + \sigma\theta\theta_x + \beta^2 \theta_{xxx} = \sigma^2 \tilde{N} = \sigma^2 \sum_{k=0}^{\infty} \sigma^k N^{(k+2)} . \qquad (2.81)$$

The arbitrary coefficients $\varPhi$ and M can be obtained by a scale transformation. The basic assumption about the right-hand side of (2.81) resembles the previous one (2.79,80). Namely, $N^{(2+k)}$ is the result of k iterations and all derivatives of the solution (this is the case for KdV) are on the order of 1 so that $|\tilde{N}| < 1$ in the interval $t \lesssim \sigma^{-1}$. We designate $\theta - \psi \equiv \delta$. Then subtracting from (2.81) the equation for ψ one gets

$$\delta_t + c\delta_x + \sigma(\delta\delta_x + \psi\delta_x + \delta\psi_x) + \beta^2 \delta_{xxx} = \sigma^2 \tilde{N}$$

or

$$\delta = -\sigma \int_0^t [\delta\delta_x(x - c(t - \tau), \sigma\tau) + (\delta\psi)_x] d\tau + \beta^2 \int_0^t \delta_{xxx} d\tau + \sigma^2 \int_0^t \tilde{N} d\tau .$$

The evolution of θ, ψ and therefore, δ is determined by $\delta(x - ct, \sigma t)$, and as follows directly, $\delta_x = \delta_\tau/c - \sigma \delta_y(\xi, y)|_{y=\sigma\tau}$. Thus, to second order

$$\sigma \int_0^t \left(\delta^2/2 + \psi\delta\right)_x d\tau = c^{-1}\sigma \left(\delta^2/2 + \psi\delta\right)\Big|_{\xi=x-ct,y=0}^{\xi=x,y=\sigma t} . \tag{2.82}$$

Transferring all the terms with σ^2, β^2 to the right side and supposing that $\delta(x, 0) = 0$ we can write

$$\delta \left[1 + \frac{\sigma}{c}\left(\delta/2 + \psi\right)\right] = \sigma^2 \int_0^t \tilde{N}' d\tau . \tag{2.83}$$

Because θ and ψ in the time interval $[0, \sigma^{-1}]$ do not exceed the unit modulus, $|\delta| \leq 2$. Estimating the modulus of (2.83) we have

$$|\delta| = \frac{\sigma^2 \int_0^t |\tilde{N}'| d\tau}{\left|1 + \frac{\sigma}{c}(\delta/2 + \psi)\right|} \leq \frac{\sigma^2 t}{1 - 2\sigma/c} . \tag{2.84}$$

The inequality (2.84) gives a stricter rule for determination of the validity interval of the KdV equation in the iteration process. The value of the amplitude σ determines the interval $t \leq (1 - 2\sigma/c)/\sigma$. Let us note that the restriction β, $\nu \leq \sigma$ is related to the condition of the regularity of the solution. The parameters β can be determined from inverse scattering methods for the KdV equation [2.71]. The definition of β is then the dimensionless reciprocal width of the narrowest possible soliton for the given initial conditions. We shall show how the given criteria work when we construct the approximate solutions for CKdV. However, in this case in (2.84) the mode velocity difference $c_n - c_m$ must replace c.

Thus the solutions must be stable. In contrast to the approach of [2.72] we suggest finding a solution not as a row, but as a series, every term of which should satisfy the more exact and complex equations. We will use nonsingular perturbation theory. Let us explain the method for CKdV.

In Sects. 2.1–4 it is shown that the evolution of the horizontal dependences in two-dimensional space at $k_x \ll 1$ is determined by the model (evolution) system of the nonlinear equations (2.71). Such a system is derived for internal and Rossby waves. In Sect. 3.3 it is demonstrated that an analogous system describes the evolution and interaction of electromagnetic waveguide modes. In Chap. 4 the ion-acoustic plasma wave interaction is considered. Note that the coefficients Φ^n_{mk} when $k = m = n$ determine the selfaction, and with different indices the coefficients are the mode interaction constants. M^n gives the velocity of dispersion and d^n_l determines the dissipation damping and intensity of "the mutual mode friction".

As is clear from this analysis, the solution of the system (2.71) can be obtained by ordinary perturbation theory only at small time. The approximate solution at large time can be derived if one chooses the KdV damping solution as the initial approximation (2.72)

$$\theta_t^{n(0)} + c_n\theta_x^{n(0)} + \sigma\Phi_{nn}^n \theta^{n(0)} + \beta^2 M^n \theta_{xxx}^{n(0)} + \tilde{\nu}d^n\theta^{n(0)} = 0 ,$$
$$\theta^{n(0)}(x,0) = \varphi^n(x) . \tag{2.85}$$

Then the solution to first order in σ will have the form

$$\theta^n(x,t) = \theta^{n(0)}(x,t) - \int_0^t \left[\sigma \sum_{m,l \neq n} \Phi_{ml}^n \theta^{m(0)} \theta_x^{l(0)} (x - c_n(t-\tau), \tau) \right.$$
$$\left. - \tilde{\nu} \sum_{n \neq l} d_l^n \theta^{l(0)}(x - c_n(t-\tau), \tau) \right] d\tau . \tag{2.86}$$

Numerical investigation of (2.86) shows that (2.85) models many characteristics of multi-mode internal waves; in particular, the decay of the initial disturbance into solitary waves, i.e., quasisolitons.

2.6 The Coupled KdV Equations

2.6.1 Quasisolitons of CKdV

Let us consider some details of the Cauchy problem for the CKdV without damping and with initially only one nonzero mode $\varphi^n = \delta_{nl}\varphi$. The evolution of the $n \neq 1$ mode is then described by (2.86). The integrals in (2.86) can be calculated approximately as in the derivation of (2.83):

$$\theta^n(x,t) = \left\{ (\sigma/2)\Phi_{ll}^n \left[\theta^{l(0)}(x,t)\right]^2 - \varphi^2(x - c_n t) \right\} / (c_l - c_n) . \tag{2.87}$$

We therefore get that without taking into account the back action, the evolution of the n mode contains both the contribution accompanying the l mode and the disturbance that propagates with its "own" velocity c_n.

The backward action is obviously the effect of the next order in the amplitude parameters σ. The resonance accumulating action of the excited mode on the mode l will be the interaction with the escorting perturbation. Neglecting, therefore, the second term in (2.87) we get the new evolution equation for the mode n:

$$\theta_t^n + c_n\theta_x^n + \sigma\Phi^n \left[\theta^n + \sigma A^n(\theta^n)^2\right] \theta_x^n + \beta^2 M^n \theta_{xxx}^n = 0 ,$$
$$A^n = \sum_{m \neq n} (\Phi_{mn}^m/2 + \Phi_{nm}^m)\Phi_{nn}^m/(c_n - c_m) . \tag{2.88}$$

Equation (2.88) can be integrated by the inverse scattering method [2.73]. The soliton solution of (2.88) is

$$\theta^n(x,t) = \left[a + \left(\frac{\sigma\Phi^n}{3(\tilde{c}_n - c_n)} - 2a \right) \cosh^2 \sqrt{\frac{\tilde{c}_n - c_n}{2\beta^2 M_n^2}}(x - \tilde{c}_n t) \right]^{-1} ,$$
$$a = \frac{\sigma\Phi^n}{3(\tilde{c}_n - c_n)} - \sqrt{\left[\frac{\sigma\Phi^n}{6(\tilde{c}_n - c_n)}\right]^2 - \frac{\sigma^2 A^n}{12(\tilde{c}_n - c_n)}} , \tag{2.89}$$

where $\tilde{c}_n$ is the velocity of the soliton propagation. When $A^n \to 0$ the mode interaction is excluded and (2.89) is transformed into the KdV soliton. The perturbation of the other modes is described by the expression

$$\theta^i(x,t) = \sigma \Phi^i_{nn} [\theta^n(x,t)]^2 / [2(c_n - c_i)] \ . \tag{2.90}$$

Equations (2.88) and (2.90) also determine the multi-soliton quasi-single-mode perturbations. It is clear that (2.88) can be the basis for the next iteration. One can be convinced that the $\sim \sigma^2$ approximation acts in the interval $\sim [0, \sigma^{-2}]$ with an error that does not exceed σ (or in the interval $[0, \sigma^{-1}]$ with an uncertainty $\sim \sigma^2$). The quasisoliton can then be interpreted as the soliton analogue for nearly integrable systems. The system of coupled KdV equations was investigated numerically. The results, which were obtained by the finite-difference scheme and by (2.86–90) coincide to within an accuracy of $\sim \sigma$. Note that if in one of the modes there is a KdV soliton while in the others $\theta^i = 0$, then such a disturbance rather quickly decays to the quasisolitons (2.89,90) and an additional perturbation on the order of $\sigma/(c_l - c_n)$. Thus, stability of a real multi-mode solitary wave can be attained only through the compensation effects of the nonlinear mode interaction by the wave form variation. The result is illustrated in Fig.2.2a,b.

In Fig.2.2c,d the decay of the initial disturbance to the quasisolitons in the two-mode case is given. The initial conditions for the modes θ^1 and θ^2 are shown by the broken lines. The result of the calculation is shown by the solid lines. The quasisolitons on the leading front are evident. The sawtooth on the second mode repeats the behavior of the analogous disturbances in the first mode [squared, as in (2.90) and therefore sharper]. The characteristic time scale of the solitons is much less than the scale of the initial disturbance and is determined by the initial amplitude and by the constants in (2.84). The validity of perturbation theory holds ($\beta \ll 1$). The initial data are for a real atmosphere in the thermospheric waveguide (auroral electrojet source) [2.25,48]. The initial mode profiles are formed at a distance of 600 km along the meridian of the source and are obtained from the solution of the linear problem. The solid lines are the solutions at a distance of 4400 km. Figures 2.2ef show the stability of quasisolitons in collisions: the broken line is the state without collision.

The decay of the large scale disturbance to waves of smaller scales is typical of solitons. Its role in the physics of the atmosphere can be important for the interpretation of spectral properties of internal waves [2.44,74]. Analogous properties appear for any waves that can be described by CKdV [2.24].

2.6.2 The Symmetry and Integrability of CKdV

Let us sum up some aspects of the mathematical CKdV [2.75]. We set $\tilde{\nu} = 0$, $\sigma = \beta = 1$. The general system has very poor symmetry. Actually if θ^s and the coordinates are transformed as $t' = Tt$, $\theta^s = \sum_k U^s_k \theta'^k + \gamma^s$, $\det U^s_k \neq 0$, $x' = X(x - \tilde{c}t)$, the invariance condition leads to the equalities

$$\sum_{k,i} U^k_n \Phi^m_{ik} \gamma^i = (Xc_n/T - c_m + \tilde{c}) U^m_n \ , \tag{2.91}$$

42 The Discrimination and Interaction

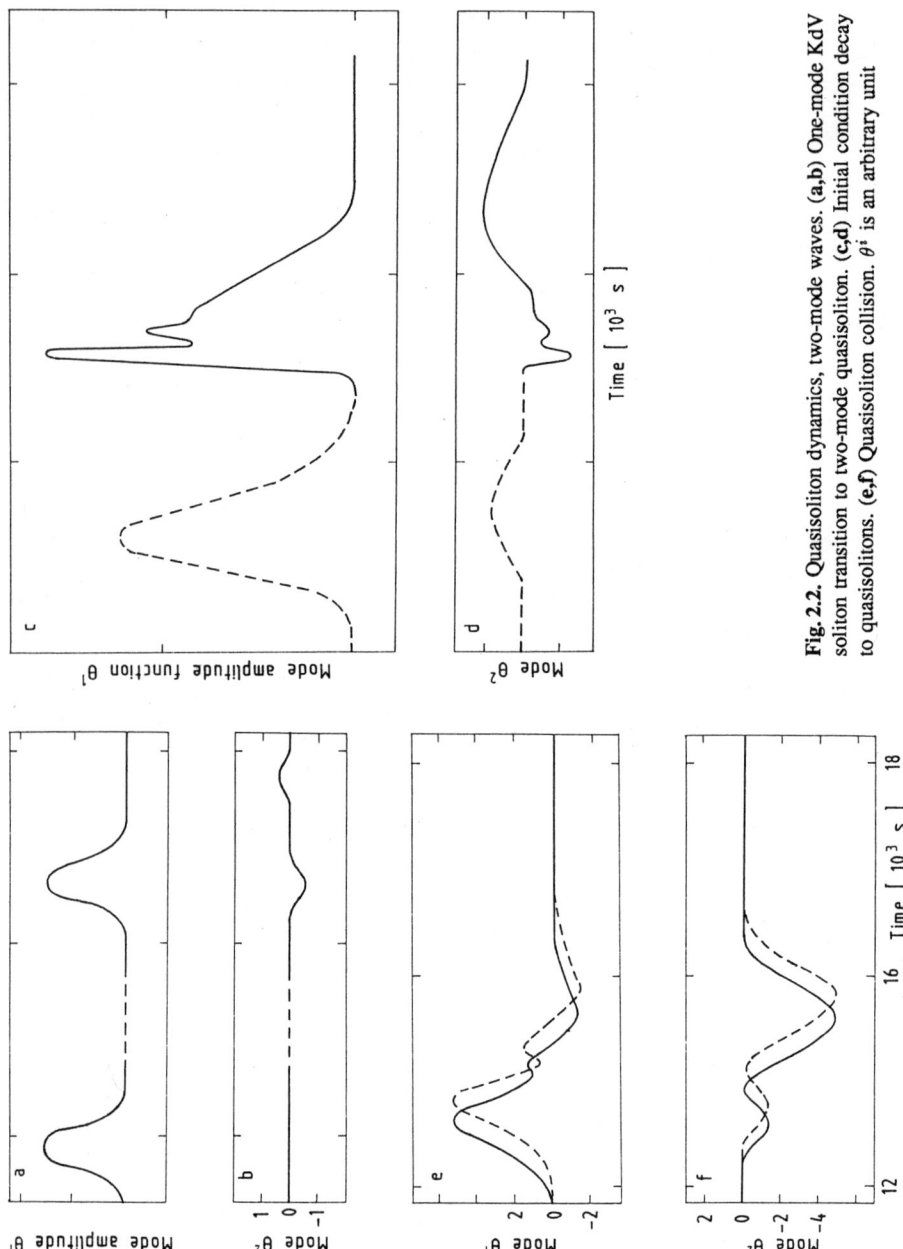

Fig. 2.2. Quasisoliton dynamics, two-mode waves. (**a,b**) One-mode KdV soliton transition to two-mode quasisoliton. (**c,d**) Initial condition decay to quasisolitons. (**e,f**) Quasisoliton collision. θ^i is an arbitrary unit

$$\sum_{s,i,k}(U_s^k)^{-1}\Phi_{il}^s U_m^i U_n^l = X\Phi_{mn}^k/T\,,\tag{2.92}$$

$$M_s U_k^s = X^3 U_k^s M_k/T\,.\tag{2.93}$$

If, as it happens in physically interesting cases, all the dispersion constants are different, then (2.93) has the consequence

$$U_k^s = \delta_k^s U_k,\ T = X^3\,.\tag{2.94}$$

The amplitude transformations are simply scale transformations and (2.91,92) are simplified

$$\sum_i \Phi_{in}^m \gamma^i = [c_n(X^{-2}-1) + \tilde{c}]\delta_n^m\,,\tag{2.95}$$

$$\Phi_{mn}^s U_m U_n = X^{-2}\Phi_{mn}^s U_s\,.\tag{2.96}$$

In the case of different Φ_{mn}^s, the n-dimensional vector γ^i should be orthogonal to $n(n-1)$ vectors Φ_{mn}^r. If, moreover, the c_s are different then there is a symmetry only with respect to the inversion $X = T = -1$. If Φ_{mn}^s and c_k are such that (2.94) is satisfied at $\gamma^i \neq 0$ then there is a continuous symmetry group. When

$$\det \Phi_{ik}^k \neq 0,\ \gamma^i = \sum_k \left(\Phi_{ik}^k\right)^{-1}[c_k(X^{-2}-1) - \tilde{c}]\,,$$

the symmetry group has two parameters. At $\det \Phi_{ik}^l = 0$, (2.95) is the condition for $X, \tilde{c}$, but free parameters γ^i may appear and (2.96) then determines the scaling relations. If $\Phi_{is}^s \neq 0$ then $U_i = X^2$. The symmetry is more complex if any of the M^n coincide. In that case the matrices U may be nondiagonal.

The possibility of L-A pair construction for the system (2.84) is also restricted. By trying to consider the system as the compatibility condition for the two linear equations

$$L\phi = (p\partial^2 + u)\phi = \lambda\phi\,,$$

$$\phi_t = A\phi = [q\partial^3 + (w-b)\partial + \partial b]\phi\,,$$

with matrix coefficients, we get the conditions

$[p, q] = 0,\ [u, q] + [p, w] = 0\,,$

$(2pw + [p, b] - 3qu)_x = 0\,,$

$(pw + 2pb - 3qu)_{xx} + [u, w] = 0\,,$

$(pb + qu)_{xxx} - wu_x - [b_x, u] = -u_t\,.$

Assuming $[u, w] = 0$ and integrating the third condition, let us express w and u as

$$w = b + p^{-1}bp - p^{-1}(s_1 x + s_2 - s_3)\,,$$

$$qu = pb + \frac{1}{3}bp - \frac{2}{3}s_2 + \frac{1}{3}s_3 - \frac{2}{3}s_1 x\,,$$

where s_k are the constant integration matrices. The desired equation has the form

$$b_t + \frac{1}{3}\tilde{b}_t - \frac{1}{3}q\tilde{b}_{xxx} - p^{-1}qN = 0,$$

$$N = [bs]b_x + \tilde{b}sb_x + \frac{1}{3}bs\tilde{b}_x + \frac{1}{3}\tilde{b}s\tilde{b}_x + b_x sb + \frac{1}{3}b_x s\tilde{b}$$
$$- \frac{1}{3}\tilde{b}b_x + Bs(b + \frac{1}{3}\tilde{b})_x + [b_x, A] - \frac{2}{3}(b + \tilde{b})q^{-1}s_1,$$

where $s = q^{-1}p$, $\tilde{b} = p^{-1}bp$, $A = (s_3 - 2s_2 - 2s_1x)/3$, $B = p^{-1}(s_3 - s_2 - s_1x)$. N contains nonlinear terms that take into account the difference between the linear velocities and the "dissipation" term which is proportional to b.

In the two-mode system with $s_1 = s_2 = 0$, choosing

$$s = cp = c\begin{pmatrix} 0 & 1 \\ 1 & 0 \end{pmatrix}, \quad q = \begin{pmatrix} 1 & 0 \\ 0 & 1 \end{pmatrix}, \quad b = \frac{1}{2}\begin{pmatrix} \theta^1 + \theta^2 & 0 \\ 0 & \theta^1 - \theta^2 \end{pmatrix}$$

we get

$$\theta_t^1 + c\theta_x^1 - \theta^1\theta_x^1 + \frac{1}{2}\theta^2\theta_x^2 - \frac{1}{4}\theta_{xxx}^1 = 0,$$

$$\theta_t^2 + \theta^1\theta_x^2 + \frac{1}{2}\theta_{xxx}^2 = 0. \tag{2.97}$$

A Walquist–Estabrook analysis leads to the same system in the general situation. For this system, first published in [2.76] and studied independently in [2.59,75], one- and two-soliton solutions have been found. The dispersion constants of the system (2.97) are opposite in sign. It is possible, as was shown in Sect.2.3, that the system describes the interaction between Rossby and internal waves. This system is undoubtedly interesting for numerical testing since it yields exact solutions.

2.6.3 The Stationary Solutions of CKdV

Let us consider the existence of stationary solutions of the general CKdV with $n = 2$. Choosing $\theta^1 = U\varphi(x - \tilde{c}t)$, $\theta^2 = V\varphi(x - \tilde{c}t)$ we introduce the designations

$$\Phi_{11}^1 = b, \; \Phi_{12}^1 = d, \; \Phi_{21}^1 = e, \; \Phi_{22}^1 = f, \; M_1 = a, \; c_1 = c,$$
$$\Phi_{22}^2 = \beta, \; \Phi_{21}^2 = \gamma, \; \Phi_{12}^2 = \varepsilon, \; \Phi_{11}^2 = \omega, \; M_2 = \alpha, \; c_2 = \sigma. \tag{2.98}$$

By means of this substitution both equations are transformed to a single KdV if the relation of the amplitudes $\xi = U/V$ and the propagation velocity satisfies the equations:

$$\tilde{c} = (\alpha c - a\sigma)/(\alpha - a),$$

$$\frac{\omega a}{\alpha}\xi^3 + \left[\frac{a}{\alpha}(\gamma + \varepsilon) - b\right]\xi^2 + \left[\frac{a\beta}{\alpha} - d - e\right]\xi - f = 0. \tag{2.99}$$

The velocity of propagation of the stationary wave in the general case is fixed and is determined by the equation constants. The condition (2.99) due to the real values of the coefficients has at least one real root. The values of the roots fix the ratio of the possible amplitudes. This difficulty of the determination of a stationary solution in a general KdV problem is due to the absence of a continuous symmetry group that is characteristic for the KdV. The Hirota-Satsuma system (2.97) has a one-parameter symmetry group. Note that [see (2.91)] $T = X^3$, $U_1 = \pm U_2 = X^{-2}$, $\gamma^2 = 0$, $\gamma^1 = c(X^{-2} - 1)/2$, $\tilde{c} = (c + \sigma)(1 - X^{-2})$. If at any value of X c is nonzero, then $\gamma^1 \neq 0$. A pure scale transform exists when $c = 0$ and indicates the possible existence of a soliton that is parameterized by amplitude [2.76]. A soliton-type stationary solution (2.98) exists with $\xi = \pm 1$, $\tilde{c} = 2c/3$.

The stationary solution of the general CKdV of the form $\theta^i = U\varphi + \lambda_i$ exists just as in the KdV theory. The propagation velocities are parameterized by λ_i.

In conclusion we can say that the derivation of a method of inverse scattering type is in all probability impossible for the general CKdV. Therefore, we aim at the development of a nonsingular perturbation theory approach that allows one to get results within a given accuracy in the time intervals in which the general CKdV is valid (Sect. 2.5). We note also that the attraction of CKdV soliton-like solutions is discussed in [2.77]. Another possible approach to coupled system investigations is based on the Painlevé test [2.78].

2.7 Nonlinear Rossby Waves on a Sphere

There are many papers devoted to nonlinear atmosphere Rossby waves [2.27, 28, 79, 80]. Most of them, however, use the β-plane approximation and even for the Rossby waves, spacial scales on the order of the earth radius and periods exceeding twenty-four hours (the azimuth periodicity) and the finiteness of the intervals of latitude variation have not been taken into account. The nonlinear wave approach allows one to describe planetary waves in real geometries. For this purpose, we shall consider the earth atmosphere, or the ocean, as a toroidal waveguide in which a large-scale wave may circulate. The difference between this new problem and that in Sect. 2.3 is the discreteness of the wave numbers and periods.

If the time is measured in days and the coordinates in units of the earth radius then the equation for the horizontal current or the geopotential ψ is [2.81]

$$[I\partial/\partial t + [\nabla\psi \times \nabla]_\perp](\Delta\psi/I + I/H - I\psi/H^2) = 0 \,. \tag{2.100}$$

The index $\perp$ denotes the radial component of the vector product. The Coriolis parameter at the colatitude v is $I = \cos v$. The angular part of the Laplace operator is denoted by Δ. The derivation of (2.100) from (2.2–5) is given in [2.6]. The equation describes barotropic modes of planetary waves.

The linearized formulation of Rossby wave propagation in a spherical coordinate system is

$$L\psi = \Delta\psi_t - I^2\psi_t/H^2 - [\nabla(I/H) \times \nabla\psi]_\perp = 0 \,,$$

with the periodicity condition on φ and the condition of finiteness of the solution in the interval $0 \leq v \leq \pi$. The Fourier method for the problem is developed by using the basis function

$$\theta_\omega^m(v) \exp i(\omega t - m\varphi), \qquad (2.101)$$

where $m = 0, \pm 1, \ldots, \omega$ are parameters of splitting which are the wavenumber and frequency. The spectrum $\omega \neq 0$ is determined from the Sturm–Liouville problem for the transverse variable ϑ

$$-\frac{1}{\sin v}(\sin v\, \theta_{\omega,v}^m)_v + \left[\frac{m^2}{\sin^2 v} + \frac{\cos^2 v}{H^2} - \frac{m}{\omega}\left(\frac{I}{H}\right)_v \frac{1}{\sin v} \right] \theta_\omega^m = 0, \qquad (2.102)$$

where θ_ω^m are in the interval $[0, \pi]$. For every m there is therefore an infinite discrete set of eigenvalues $\omega_n(m)$. Due to the positively determined operator in (2.102) the ratio $m/\omega \leq 0$. The eigen functions θ_ω^m are orthogonal and weighted according to

$$\int_0^\pi \left(\frac{I}{H}\right)_v \theta_\omega^m(v) \theta_{\omega'}^m dv = 0$$

when $\omega \neq \omega'$.

Let us pay attention to the special case $m = 0$. In this case the time dependence in the linear approximation is absent. The specific state, a zonal flow, can be considered as the ground state of the atmosphere. Its variations in latitude are given by θ^0 and its time evolution is determined by nonlinear interaction with other modes and sources in the atmospheric gas.

The meridional variations of the background temperature (height H) in the layer nearest to the earth usually do not exceed 100 K. Therefore the function $H(v) = H_0 + H_1(v)$ does not vary much. Then in the simplest version of the theory we can neglect H_1 since it is small compared with H_0. In this case we get from (2.102), omitting indices and putting $\cos v = \xi$,

$$-[\theta_\xi(1-\xi^2)]_\xi + \left(\frac{m^2}{1-\xi^2} + \frac{\xi^2}{H_0^2} + \frac{m}{\omega}\frac{1}{H_0}\right)\theta = 0.$$

This equation coincides with that for the spheroidal functions [2.82].

We shall seek the solution of the nonlinear problem in the form of a multi-wave system which is a superposition of the mode contributions (2.101) with coefficients that depend on the "slow" variables φ, t. The dispersion relation is formally the same as (2.34):

$$\omega = m/H(k_v^2 + m^2 + 1/H^2).$$

Due to the discrete spectrum, m is not small compared with k_v and the strong dispersion case occurs. Thus we put

$$\psi = \sum_{m,\omega} \theta_\omega^m(v) A_\omega^m(\beta\varphi, \beta t)\, e^{i(\omega t - m\varphi)} + \text{c.c.}. \qquad (2.103)$$

The interaction which appears due to nonlinearity is taken into account through the substitution of (2.103) into the nonlinear terms of (2.100)

$$N = -\frac{1}{\sin v}\left[\psi_v\left(\frac{\Delta\psi}{I} - \frac{I\psi}{H^2}\right)_\varphi - \psi_\varphi\left(\frac{\Delta\psi}{I} - \frac{I\psi}{H^2}\right)_v\right].$$

The result is put in (2.100) with the contribution to first order in $\sim \beta$. The basic three-wave interaction is the main resonance for the quadratic nonlinearity. Denoting

$$A_\omega^m e^{i(\omega t - m\varphi)} \pm A_\omega^{*m} e^{-i(\omega t - m\varphi)} \equiv A_k^\pm$$

and using linear relations between variables in the nonlinear expression (Sect. 2.5), we can write the full equation

$$L\psi = N = -\frac{i}{I}\sum_{m',\omega'}\frac{m}{\sin v}\left(\frac{m}{\omega}\theta_\omega^m\theta_{\omega',v}^{m'}A_{k'}^+A_k^-\right.$$
$$\left.-\frac{m'}{\omega'}\theta_\omega^m\theta_{\omega'}^{m'}A_{k'}^-A_k^+\right)\left(\frac{I}{H}\right)_v. \tag{2.104}$$

Let us project the equation onto the mode basis. To do this we integrate (2.104) with the function $\sin v\,\theta_{\omega_1}^{m_1}$ and change the indices:

$$\int_0^\pi \sin v\,\theta_{\omega_1}^{m_1} L\psi\,dv = \sum_{m\omega}\left[\int_0^\pi \theta_{\omega_1}^{m_1}\left(\frac{I}{H}\right)_v \theta_\omega^m dv\left(\frac{m}{\omega}A_t + A_\varphi\right)\right.$$

$$\left. + 2m\omega\int_0^\pi \theta_{\omega_1}^{m_1}\theta_\omega^m\frac{dv}{\sin v}A_\varphi\right]e^{i(\omega t - m\varphi)} + \text{c.c.}$$

$$= i\sum_{\substack{m_2 m_3\\ \omega_2\omega_3}}\int_0^\pi\frac{m_2}{I}\theta_{\omega_1}^{m_1}\theta_{\omega_2}^{m_2}\left[W(\theta_{\omega_3}^{m_3})_v\left(\frac{m_3}{\omega_3} - \frac{m_2}{\omega_2}\right)\right.$$

$$\left. + \frac{m_3}{\omega_3}W_v\theta_{\omega_3}^{m_3}\right]dv A_{k_3}^+ A_{k_2}^- = i\sum_{\substack{m_2 m_3\\ \omega_2\omega_3}}N_{\omega_1\omega_2\omega_3}^{m_1 m_2 m_3}A_{k_3}^+A_{k_2}^-,$$

where $W = (I/H)_v/\sin v$.

The condition of wave synchronicity $m_1 = m_2 + m_3$ determines sets of frequencies from the spectra of (2.102). The resonance and the dominant interaction is reached at $\omega_1 = \omega_2 + \omega_3$.

The values of frequencies were calculated for $m = 0$ to 10 for periods of up to 60 days. The computer program which finds the resonance triads to a chosen degree of accuracy and the results of the calculations are given in a thesis (*Yu.Brezhnev*, Kaliningrad State University, 1985). Examples of resonance triads and periods in intervals of 23 days are given in Tables 2.2, 3. In Table 2.3 the triads are numbered by a pair of indices (m, l) where m is a wave number and l is the index of the eigen frequency. The accuracy of the frequencies has been checked against [2.83]. The values of the nonlinear constants allow one to estimate the amount of energy exchanged between the resonance modes.

Unlike the internal waves interaction (their nonlinear constants decrease with the mode number, see Table 2.1), the Rossby waves behave in a complex manner. Nevertheless, it is possible to distinguish the 12–day oscillations observed in nature [2.27]. As it is seen from Table 2.3 the nonlinear constants of modes of nearly equal periods are considerably larger.

Table 2.2. Planetary wave periods [days]. m – wave number, l – eigenfrequency

l \ m	1	2	3	4	5	6
1	1.65	4.52	7.79	11.85	16.87	22.88
2		1.75	3.60	5.76	8.33	11.37
3			2.13	3.67	5.44	7.50
4				2.58	3.96	5.54
5					3.06	4.35
6						3.54

Table 2.3. Three-wave interaction

Difference	Triads (m, l)			Periods [days]			Nonlinear constants			Propagation velocities		
$\omega_1 + \omega_2 - \omega_3$	3	2	1	3	2	1	3	2	1	3	2	1
−1.4	(4,3)	(8,2)	(2,1)	3.67	18.9	4.5	0.21	0.014	0.3	0.21	0.15	0.99
1.7	(6,3)	(6,2)	(6,1)	7.50	11.4	22.9	1.1	0.73	0.035	0.06	0.19	0.31
0.3	(8,3)	(8,2)	(8,1)	12.6	18.9	37.9	−1.0	−1.1	−0.06	0.09	0.15	0.20
0.7	(8,3)	(10,2)	(6,1)	12.6	28.4	22.9	−1.8	−0.3	−3.0	0.09	0.11	0.31
−2.6	(7,4)	(8,3)	(5,1)	7.34	12.6	16.9	−0.54	−0.24	−0.51	0.007	0.09	0.40
1.9	(7,5)	(9,4)	(4,1)	5.81	11.7	11.9	−3.3	−0.98	−4.7	0.10	0.04	0.52

For some triads, redenoting $X = A^{\omega_1}_{m_1}$, $Y = A^{\omega_2}_{m_2}$, $Z = A^{\omega_3}_{m_3}$, let us write a three-wave system in standard form

$$X_t + v_1 X_\varphi = i\beta_1 Y Z ,$$
$$Y_t + v_2 Y_\varphi = i\beta_2 \bar{X} Z ,$$
$$Z_t + v_3 Z_\varphi = i\beta_3 \bar{Y} X , \qquad (2.105)$$

where

$$v_i = \left[(1 + 2m_i\omega_i) \int_0^\pi (\theta^{m_i}_{\omega_i})^2 dv / \sin v\right] \omega_i / m_i c_i ,$$

$$c_i = \int_0^\pi (\theta^{m_i}_{\omega_i})^2 \left(\frac{I}{H}\right)_v dv ,$$

and where

$$\beta_1 = \omega_1 \left(N^{m_1 m_2 m_3}_{\omega_1 \omega_2 \omega_3} + N^{m_1 m_3 m_2}_{\omega_1 \omega_3 \omega_2}\right) / c_1 m_1 ,$$

$$\beta_2 = \omega_2 \left(N^{m_2 m_1 m_3}_{\omega_2 \omega_1 \omega_3} - N^{m_2 m_3 m_1}_{\omega_2 \omega_3 \omega_1}\right) / c_2 m_2 ,$$

$$\beta_3 = \omega_3 \left(N^{m_3 m_2 m_1}_{\omega_3 \omega_2 \omega_1} - N^{m_3 m_1 m_2}_{\omega_3 \omega_1 \omega_2}\right) / c_3 m_3 .$$

The system (2.105) describes the interaction of three modes with the complex amplitude functions X, Y, Z. The type of process depends on the values of the coefficients [2.19]. The constants β_i are listed in Table 2.3; the calculations were made on the basis of normal functions θ_ω^m. The signs of the constants determine the solution (2.105) behavior, for example, whether "the explosive instability" exists. The classification of the three-wave interaction, however, is based on the case of equal group velocities. The different propagation velocities and the localization of the wave packet change the lifetime and the outcome of the interaction. A full description of the interaction processes of localized packets is based on the inverse problem method and yields restricted solutions [2.73].

This problem is related to the class of periodical problems. The functions to be found should be periodic in φ. Some aspects can be analyzed for the specific case $v_2 \approx v_3 \equiv u$, $v_1 \equiv v$. If the coefficients β_i are of the same sign one can get a system with unit nonlinear constants by the transformation of scale amplitudes. Then, introducing new variables $\xi = (x - vt)/(u - v)$, $\tau = (x - ut)(v - u)$, $S = |Y'|^2 + |Z'|^2$ we get the selfinduced transparency equations. For a real amplitude X and $|S| \leq 1$, the obtained equations are transformed [2.7] to the sine-Gordon (SG) equation

$$\psi_{\xi\tau} = \sin\psi ,$$

where $\psi_\xi = X'$, $S = \cos\psi$. In considering the Rossby wave mode and the zonal flow interaction in a laboratory device [2.22], the authors of [2.17] have also obtained an SG equation. In the same work the solutions of SG in the form of a Lamb ansatz and the corresponding nonlinear dispersion relations were investigated. The example of the finite-gap real and periodic solution is described in [2.84]. The solution behavior as a function of time is regular and demonstrates a periodical energy change between modes. A review of finite-gap solutions for the SG equation is given in [2.85]. Integrable multiple three-wave interaction equations were considered in [2.86,87].

3. Interaction of Modes in an Electromagnetic Field Waveguide

Nonlinear electromagnetic modes in waveguides (Sect. 3.1) interact and disperse (Sects. 3.1,2) similarly to hydrodynamic ones but the origin of interaction is related to the microscopic properties of the medium. In this chapter, it is shown how the nonlinear constants may be represented and calculated in terms of a density matrix (Sect. 3.3). The boundary conditions in metal or dielectric waveguides leads to the physically possible finite-mode case. The nonlocal dispersion operator may appear due to boundary matching of solutions for dielectric waveguides (dielectric slabs, rods and fibers). The derivation of this operator is given in Sect. 3.4, where the generalization of the projection procedure for the case of noncommuting matrix elements of the basic linear evolution operator is given.

3.1 Electromagnetic Fields in Dielectric Waveguides

3.1.1 Basic Equations and Boundary Conditions

Electromagnetic wave propagation in continuous media is described by the Maxwell equations and constitutive equations which relate the polarization and magnetization vectors to the electromagnetic field [3.1]. Because we are concerned with nonlinear phenomena we choose as the equation for the magnetic field H, restricting ourselves to magnetic interactions,

$$-\Delta H + \varepsilon\mu H_{tt}/c^2 = -\varepsilon M_{tt}/c^2 + \nabla \operatorname{div} M/\mu , \qquad (3.1)$$

where ε is the dielectric constant and magnetic permeability is divided into two contributions: the nonresonant μ and the resonant parts. The remaining contributions are taken into account in the magnetization vector $M = N\mu$. The dependence of the projection of the magnetic momentum onto the magnetic field determines the degree of nonlinearity of the medium. The dispersion of the electromagnetic waves is determined mainly by the waveguide propagation. The velocity of propagation depends on the wave mode number and for every mode function $\omega_n(k_x)$ it is nonlinear [3.2]. Let us trace all the steps of the derivation of the model evolution equation beginning at the microscopic level.

In the quantum theory the evolution of the magnetic moment is determined by the dependence of the density matrix ϱ on time. Thus $\mu = Sp\varrho\hat{\mu}$, where $\hat{\mu}$ is the vector operator of the magnetic moment. This equation is key for the connection between the magnetization vector M and the magnetic field H, i.e., for the form of the constitutive equation. If I_k are the components of the spin operator then the spin contribution to a magnetic moment operator is

$$\hat{\mu}_k = \mu_0 \mu_s I_k , \qquad (3.2)$$

where $\mu_0 \mu_s \equiv \gamma$ is the gyromagnetic ratio, μ_0 is the Bohr magneton. For the evolution of the density matrix it is convenient to use for the matrix elements

$$\dot{\varrho}_{sk} = (E_s - E_k)\varrho_{sk}/i\hbar + [V,\varrho]_{sk} - (\varrho_{sk} - \delta_{sk}\varrho_s^0)/T_{sk} . \qquad (3.3)$$

The last addend in (3.3) introduces, phenomenologically, a relaxation with a characteristic transition time T_{sk}. E_i are energy levels of the unperturbed quantum system $\varrho(t=0) \equiv \varrho^0$. The dependence on the field components is localized in the interaction operator V. The simplest operator form for the interaction between the nuclei and the magnetic field is

$$V = -\gamma(H, I) , \qquad (3.4)$$

i.e, V depends on the magnetic field components H_k by a linear function. The dependence of $\varrho(H_k)$, and therefore of $\mu_s(H_k)$, is nonlinear because of (3.3). If we, as in previous discussions, restrict ourselves to weak nonlinearity, then we can proceed with second order perturbation theory when seeking $\varrho(H_k)$. Then $\varrho = \varrho^{(1)} + \varrho^{(2)}$, the superscript in parentheses is the order of the perturbation parameter. It can be easily seen that in the absence of relaxation

$$\varrho^{(1)} = \int_{-\infty}^{t} dt_1 [V(t_1), \varrho^0]/i\hbar ,$$

$$\varrho^{(2)} = \int_{-\infty}^{t} dt_1 \int_{-\infty}^{t_1} dt_2 V(t_1) [V(t_2), \varrho^0] /(i\hbar)^2 .$$

Therefore the vector $M^{(2)} = N Sp \varrho^{(2)} \hat{\mu}$ entering (3.1) is a quadratic form of the magnetic field components H_k. Thus, in the nonlinear problem the components of (3.1) are linked. The details of the calculation for this example will be shown in Sect. 3.3. The linear part of the resonance contribution depends on $\varrho^{(1)}$ and is equal to

$$\mu^{(1)} = Sp\hat{\mu}\varrho^{(1)} = \int_{-\infty}^{t} dt_1 Sp\hat{\mu}[V(t_1), \varrho^0] .$$

For simplicity we suppose that the medium is isotropic and include the magnetic moment into μ. Denoting

$$\Box = \frac{\varepsilon\mu}{c^2}\frac{\partial^2}{\partial t^2} - \Delta ,$$

we get

$$\Box H_i = -\frac{\varepsilon}{c^2} M_{itt}^{(2)}(H_1, H_2, H_3) + \frac{1}{\mu}\nabla_i \operatorname{div} M^{(2)} = N_i . \tag{3.5}$$

The relations between the components of the field vectors inside nonlinear terms will be determined by the linearized Maxwell equations. Just as in Sect. 2.5, we neglect the contributions of third order. We shall write the linear Maxwell equations in the form

$$\operatorname{div} \boldsymbol{E} = 0 , \quad \operatorname{rot} \boldsymbol{E} = -\frac{\mu}{c}\frac{\partial \boldsymbol{H}}{\partial t} ,$$

$$\operatorname{rot} \boldsymbol{H} = \frac{\varepsilon}{c}\frac{\partial \boldsymbol{E}}{\partial t} , \quad \operatorname{div} \boldsymbol{H} = 0 . \tag{3.6}$$

An important role in the study of waveguide propagation is played by boundary conditions. It is well known [3.1,3] that due to the equations themselves, the normal component of the magnetic induction vector $\boldsymbol{B}$ and the tangential component of $\boldsymbol{E}$ are continuous on the dividing surface

$$B_{1n} = B_{2n} , \quad E_{1t} = E_{2t} . \tag{3.7}$$

The normal components of $\boldsymbol{D}$ and $[\boldsymbol{n} \times (\boldsymbol{B}_2 - \boldsymbol{B}_1)]$ have discontinuities where the surface densities of the charge and the current of the external sources are

$$\sigma = \frac{1}{4\pi}\lim_{\varepsilon \to 0}\int_{l_\varepsilon} dl \operatorname{div}[\boldsymbol{n}\times[\boldsymbol{D}\times\boldsymbol{n}]] ,$$

$$j = \frac{1}{4\pi}\lim_{\varepsilon \to 0}\int_{l_\varepsilon} dl \frac{\partial \boldsymbol{D}}{\partial t} . \tag{3.8}$$

Integration is performed over the contours close to the surface.

The conditions (2.7,8) are obtained with linear equations. If the constitutive equations are nonlinear, then the relations on the surface may be nonlinear too. In the problems considered, volume effects exceed surface effects, therefore weak surface nonlinearity is neglected.

3.1.2 Waveguide Modes

Electromagnetic wave propagation in waveguides [3.4] is very interesting for the study of nonlinear mode interactions. Due to the existence of critical frequencies one can choose a configuration and dimension of the waveguide so that only a few modes can exist in the given wave vector interval. In linear theory there is a corollary to the existence of critical frequencies; it is interesting to test whether the excitation of forbidden modes may be due to intermode interaction.

Let us consider a rectangular waveguide with conducting walls of height a and width b. The axis z is directed vertically, y is along the base and x is along the waveguide axis (the direction of the wave propagation). Boundary conditions (3.7) in this chosen geometry of the waveguide give

$$E_{x,y}\big|_{z=0,b} = 0 , \quad E_{x,z}\big|_{y=0,a} = 0 ,$$

$$H_z\big|_{z=0,b} = 0 , \quad H_y\big|_{y=0,a} = 0 . \tag{3.9}$$

It is convenient to introduce basis functions beginning from E_x, since on the entire perimeter of the waveguide this projection of the electric field should be zero due to (3.9). Therefore, solving the simplest spectral problem we get

$$E_x = \sum_{n,m} \varphi^{nm}(x,t) \sin \frac{m\pi}{a} y \sin \frac{n\pi}{b} z \,. \tag{3.10}$$

E_y and E_z are easily chosen by taking into account that div $\boldsymbol{E} = 0$:

$$E_y = \sum_{n,m} \phi^{nm}(x,t) \cos \frac{m\pi}{a} y \sin \frac{n\pi}{b} z \,,$$

$$E_z = \sum_{n,m} \eta^{nm}(x,t) \sin \frac{m\pi}{a} y \cos \frac{n\pi}{b} z \,. \tag{3.11}$$

Further, using the remaining Maxwell equation, namely rot $\boldsymbol{E} = -\mu \boldsymbol{H}_t/c$, we write

$$H_x = \sum_{n,m} \alpha^{nm}(x,t) \cos \frac{m\pi}{a} y \cos \frac{n\pi}{b} z \,,$$

$$H_y = \sum_{n,m} \beta^{nm}(x,t) \sin \frac{m\pi}{a} y \cos \frac{n\pi}{b} z \,,$$

$$H_z = \sum_{n,m} \gamma^{nm}(x,t) \cos \frac{m\pi}{a} y \sin \frac{n\pi}{b} z \,. \tag{3.12}$$

The substitution into the complete Maxwell system gives (omitting the mode indices)

$$\varphi_x - \frac{m\pi}{a}\phi - \frac{n\pi}{b}\eta = 0 \,,$$

$$-\frac{\mu}{c}\alpha_t = \frac{m\pi}{a}\eta - \frac{n\pi}{b}\phi \,, \tag{3.13}$$

$$-\frac{\mu}{c}\beta_t = \frac{n\pi}{b}\varphi - \eta_x \,,$$

$$-\frac{\mu}{c}\gamma_t = \phi_x - \frac{m\pi}{a}\varphi \,, \tag{3.14}$$

$$\alpha_x + \frac{m\pi}{a}\beta + \frac{n\pi}{b}\gamma = 0 \,,$$

$$\frac{\varepsilon}{c}\varphi_t = -\frac{m\pi}{a}\gamma + \frac{n\pi}{b}\beta \,, \tag{3.15}$$

$$\frac{\varepsilon}{c}\phi_t = -\frac{n\pi}{b}\alpha - \gamma_x \,,$$

$$\frac{\varepsilon}{c}\eta_t = \beta_x + \frac{m\pi}{b}\alpha \,. \tag{3.16}$$

It is not difficult to verify that (3.14) and (3.16) follow from the others and, moreover, that all the coefficient functions satisfy similar equations

$$\alpha_{xx} - \frac{\mu\varepsilon}{c^2}\alpha_{tt} = \kappa^2 \alpha,$$

$$\kappa^2 = \pi^2(m^2b^2 + n^2a^2)/a^2b^2 \,. \tag{3.17}$$

The minimum value of κ determines the limit of the waveguide penetrability because the electromagnetic wave $\sim \alpha = \alpha_0 \exp[i(\omega t - kx)]$ can exist only if the parallel component of the wave vector is a real number, and due to (3.17) $k^2 = \omega^2 \varepsilon \mu/c^2 - \kappa^2$. In other words, the frequency ω should be more than $\kappa^2 c^2/\varepsilon\mu$. It is namely this condition, along with the discrete nature of the spectrum of κ which allows a finite number of modes to be excited by varying the source frequency or the waveguide dimensions.

Now we write (3.13–16) in the form of (2.29) excluding η and γ from (3.13,15).

$$\frac{\varphi_t}{c} = \frac{mb}{na\varepsilon}\alpha_x + \frac{\kappa^2 b}{n\pi\varepsilon}\beta ,$$

$$\frac{\phi_t}{c} = -\frac{n\pi}{\varepsilon b}\alpha + \frac{b}{n\pi\varepsilon}\alpha_{xx} + \frac{mb}{\varepsilon na}\beta_x ,$$

$$\frac{\alpha_t}{c} = -\frac{mb}{na\mu}\varphi_x + \frac{\kappa^2 b}{n\pi\mu}\phi , \qquad (3.18)$$

$$\frac{\beta_t}{c} = -\frac{n\pi}{\mu b}\varphi + \frac{b}{n\pi\mu}\varphi_{xx} - \frac{mb}{na\mu}\phi_x .$$

If $\Phi^T = (\varphi, \phi, \alpha, \beta)$, then $\Phi_t/c = L\Phi$. The Fourier transformation on x gives

$$L(ik) = \begin{pmatrix} 0 & 0 & s & p \\ 0 & 0 & q & r \\ -s\varepsilon/\mu & p\varepsilon/\mu & 0 & 0 \\ q\varepsilon/\mu & -r\varepsilon/\mu & 0 & 0 \end{pmatrix}, \qquad (3.19)$$

where $s = ikmb/\varepsilon na$, $p = \kappa^2 b/\pi n\varepsilon$, $q = \left(-n^2\pi^2 + b^2 k^2\right)/bn\pi\varepsilon$, $r = -ikmb/\varepsilon na$. The rank of the matrix L appears to be equal to three, therefore a set of solutions is divided into two independent classes.

In the waveguide propagation theory two types of solutions of (3.18) are usually discussed. If $H_x = 0$ or $\alpha = 0$ we get a field of the type "E" or a transverse–magnetic wave (TM). Denoting it by the index e let us write the equations relating the components

$$\beta^e = \frac{n\pi\varepsilon}{c\kappa^2 b}\varphi_t^e, \qquad \phi^e = \frac{\pi a}{m\kappa^2}\varphi_x^e ,$$

$$\gamma^e = \frac{m\pi\varepsilon}{c\kappa^2 a}\varphi_t^e, \qquad \eta^e = \frac{na^2\pi}{m^2\kappa^2 b}\varphi_x^e .$$

Choosing $E_x = 0$ or $\varphi = 0$ we shall derive the field of an "H"–type transverse–electric wave (TE). The corresponding component relationships are (indices h)

$$\beta^h = -\frac{m\pi}{\kappa^2 a}\alpha_x^h, \qquad \phi^h = -\frac{n\pi\mu}{c\kappa^2 a}\alpha_t^h ,$$

$$\gamma^h = -\frac{n\pi}{\kappa^2 b}\alpha_x^h, \qquad \eta^h = \frac{m\pi\mu}{c\kappa^2 a}\alpha_t^h . \qquad (3.20)$$

Since α and φ satisfy the second-order equation (3.17) they are determined by two initial ($t = 0$) or boundary ($x = 0$) conditions and both contain contributions that correspond to oppositely directed waves. This classification is supported by the dispersion equation

$$\det[L(-i\boldsymbol{k}) - i\omega] = 0 \tag{3.21}$$

which, after the substitution of (3.19), leads to two equations of second order.

The general form of the magnetic field expansion includes contributions of both types, since in nonlinear evolution dynamics, due to mode interaction, an additional wave field can be generated. Therefore, only the x-component contains the magnetic terms:

$$H_x = \sum_{mn} \alpha_h^{mn} \cos\frac{m\pi}{a}y \cos\frac{n\pi}{b}z \equiv \sum \alpha^{mn} C^{mn} . \tag{3.22}$$

The functions of the mode basis are denoted by C^{mn} and the index h is omitted. Since

$$\sin\frac{m\pi}{a}y\cos\frac{n\pi}{b}z = -\frac{a}{m\pi}C_y^{mn}; \quad \frac{b}{n\pi}C_z^{mn} = -\cos\frac{m\pi}{a}y\sin\frac{n\pi}{b}z ,$$

the other components are

$$H_y = -\sum_{mn}\frac{1}{\kappa^2}\left(\frac{na\varepsilon}{mbc}\varphi_t^{mn} - \alpha_x^{mn}\right)C_y^{mn} , \tag{3.23}$$

$$H_z = -\sum_{mn}\frac{1}{\kappa^2}\left(\frac{mb\,\varepsilon}{na\,c}\varphi_t^{mn} - \alpha_x^{mn}\right)C_z^{mn} . \tag{3.24}$$

3.2 Equations of Mode Interaction

The expression for the mode series obtained in Sect. 3.1 for the magnetic field can now be put into (3.5). Afterwards, the equalities are multiplied by their corresponding mode functions. Using the orthogonality of the sine and cosine functions in the interval $[0, \pi]$ and taking into account that $\Delta C^{mn} = -\kappa^2 C^{mn}$, we get

$$(\Box_1 + \kappa^2)\alpha^{mn} = \frac{(C^{mn}, N_1)}{\|C^{mn}\|} , \tag{3.25}$$

$$(\Box_1 + \kappa^2)\left(-\frac{na\varepsilon}{mbc}\varphi_t^{mn} + \alpha_x^{mn}\right) = \frac{4\kappa^2(C_y^{mn}, N_2)}{m\pi b} , \tag{3.26}$$

$$(\Box_1 + \kappa^2)\left(-\frac{mb\varepsilon}{nac}\varphi_t^{mn} + \alpha_x^{mn}\right) = \frac{4\kappa^2(C_z^{mn}, N_3)}{n\pi a} . \tag{3.27}$$

Here $\Box_1 \equiv (\varepsilon\mu/c^2)(\partial^2/\partial t^2) - (\partial^2/\partial x^2)$. Every mode, as follows from the dispersion relation (3.21) and directly from (3.25–27), contains contributions of the "right" and "left" waves. More exactly, "E" and "H" waves are divided into such components. We denote them correspondingly by $\alpha = \alpha^{\Pi} + \alpha^{\Lambda}$, $\varphi = \varphi^{\Pi} + \varphi^{\Lambda}$. See also Chap. 1.

Let us discuss more thoroughly the properties of the dispersion relation. Its simplest form follows from the operator form

$$\Box_1 + \kappa^2 (c_0 \equiv c/\sqrt{\varepsilon\mu}) ,$$
$$\omega = c_0 \sqrt{k^2 + \kappa^2} . \tag{3.28}$$

If we denote $k_0 = \omega/c_0$ and suppose that the excitation of the waveguide is given by the frequency then (3.28) should be represented as

$$\omega^2 = \frac{\kappa^2 c_0^2}{1 - (k/k_0)^2} .$$

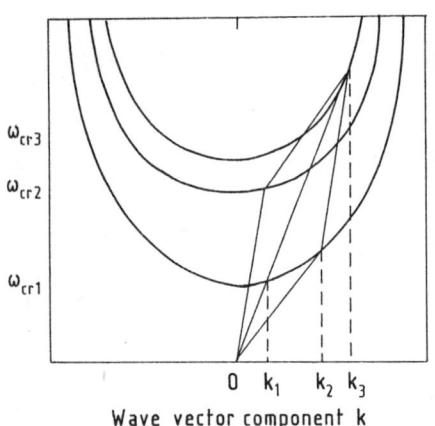

Fig. 3.1. Dispersion curves and three-wave interaction diagram for electromagnetic waveguide propagation

The dispersion curves are shown in Fig. 3.1. The critical frequencies $\omega_{kp_i} = c_0 \kappa_{mn}$ are numbered in increasing order. The dispersion that is determined by (3.28) is strong [3.2,5]. The left and right waves in this case are given in the form of wave trains. The group velocity modulus is equal to

$$\omega'(k) = \frac{c}{\sqrt{\varepsilon\mu}} \frac{k}{\sqrt{\kappa^2 + k^2}} \equiv v = \frac{c_0^2 k}{\omega} .$$

The width of the wavepacket is characterized by the dispersion parameter β. The condition $\beta \sim \Delta k/2\kappa \ll 1$ should be valid. Let us consider the mode mn and suppose for generality that all the boundary conditions (at $x = 0$) represent the trains of oscillations at different frequencies:

$$\alpha = X^{\Pi}(x - v_1 t)e_1^- + X^{\Lambda}(x + v_2 t)e_2^+ ,$$
$$\varphi = Y^{\Pi}(x - v_3 t)e_3^- + Y^{\Lambda}(x + v_4 t)e_4^+ ,$$

where $e_s^{\pm} = \exp[i(\omega_s t \pm k_s t)]$, $v_s(\omega_s)$ is the group velocity modulus. The next step now is a derivation of the equations for the amplitudes X, Y. As earlier, this procedure involves the introduction of small parameters of nonlinearity σ and dispersion β. The dynamical wave functions then have the typical form

$$f = \sigma X(\xi, \sigma t)e^{i(\omega t - kx)}, \quad \xi = (x - vt)\beta. \tag{3.29}$$

If the relation between small parameters is $\sigma \sim \beta^2$, then to second order in σ we can write ($\tau = \sigma t$)

$$\left(\frac{1}{\kappa^2}\Box_1 + 1\right)f = \frac{2i\omega\sigma^2}{\kappa^2 c_0^2}\left[X_\tau + \frac{\beta^2}{\sigma}\frac{v^2 - c_0^2}{2i\omega}X_{\xi\xi}\right]e^-. \tag{3.30}$$

Let us calculate the "kinematic" constants in the model evolution system. Suppose that the monodirectional wave is initially excited and neglect the interaction with generated modes at appropriate times. It is then possible to write the equation with the Schrödinger linear operator

$$iX_\tau + \frac{\beta^2}{\sigma}\frac{v^2 - c_0^2}{2\omega}X_{\xi\xi} = \frac{c_0^2 e^{-1}(C^{mn}, N_1)}{2\omega\sigma^2\|C^{mn}\|^2}. \tag{3.31}$$

Here we have used (3.25,30). If X is related to H_x by the projection of the magnetic field via (3.29), i.e., the magnetic wave is excited, then, as follows from (3.20), all components of the magnetic field of the mode are nonzero. N^1 is their quadratic form. At the same time the resonant term is absent, as follows from (3.20), and there is no accumulation. Such effects appear in the next higher orders of iteration or if the mode interaction is taken into account. The second iteration gives a cubic nonlinearity and (3.31) becomes the nonlinear Schrödinger equation (NS).

The wave modes interaction yields a resonant contribution already in the first order. Therefore it is sufficient to consider a one-scale approximation

$$f = X(\beta x, \beta t)e^{i(\omega t - kx)},$$

$$\left(\frac{1}{\kappa^2}\Box_1 + 1\right)f = \frac{2i\beta\omega}{\kappa^2 c_0^2}\left[X_\tau - \frac{c_0^2 k}{\omega}X_\xi\right]. \tag{3.32}$$

Let us consider (3.22–27,31). First write the expressions for the projections of the magnetic field; they are to be put into the right-hand side:

$$H_x = \sum_{m,n}(X^{\Pi mn}e_1 + X^{\Lambda nm}e_2)C^{mn}, \tag{3.33}$$

$$H_y = -\sum_{m,n} i\left\{\frac{na\varepsilon}{mbc}(\omega_3 Y^{\Pi mn}e_3 + \omega_1 Y^{\Lambda mn}e_4)\right.$$

$$\left. - (k_1 X^{\Pi mn}e_1 - k_2 X^{\Lambda mn}e_2)\right\}C_y^{mn}\kappa^{-2}, \tag{3.34}$$

$$H_z = -\sum_{m,n} i \left\{ \frac{mb\varepsilon}{nac} \left(\omega_3 Y^{\Pi mn} e_3 + \omega_4 Y^{\Lambda mn} e_4 \right) \right.$$
$$\left. - \left(k_1 X^{\Pi mn} e_1 - k_2 X^{\Lambda mn} e_2 \right) \right\} C_z^{mn} \kappa^{-2}. \tag{3.35}$$

Now using the linear dependence of the exponents e_s and assuming that the frequency intervals of the wave trains do not intersect, we can write the equations for multi-wave interacting modes as in Sect. 2.7. We set the parameter $\beta = 1$:

$$X_t^{\Lambda mn} - v_2 X_x^{\Lambda mn} = \frac{\kappa^2 c_0^2 e_2^{-1}}{2 i \omega_2 \| C^{mn} \|^2} (C^{mn}, N^1), \tag{3.36}$$

$$Y_t^{\Pi mn} + v_3 Y_x^{\Pi mn} = \frac{\pi \kappa^2 c_0^2 cbe_3^{-1}}{i \omega_3^2 \varepsilon n \| C^{mn} \|^2} (C_y^{mn}, N^2). \tag{3.37}$$

The equations for X^Π, Y^Λ are analogous. The nonlinear resonance and therefore the strength of the interaction depends on the structure of N^s which in turn is determined by the Hamiltonian and the initial value of the density matrix (Sect. 3.1). To understand this, we need to know the details of the physical assumptions.

In addition, for the existence of an effective n-wave interaction, the frequencies ω_i^{mn} and vectors k_i^{mi} should obey special conditions. These are illustrated in Fig. 3.1 which shows a three-wave resonance and synchronicity. In Appendix 4 these conditions are given in algebraic form.

3.3 Three-Wave Interaction of Magnetic Modes in an Extended System of Nuclear Spins

In this section we discuss the theory of propagation of electromagnetic waves in a medium whose nonlinearity is due to the peculiarities of the time dependence of the nuclear spins. It is assumed that the field frequency lies in the NQR-NMR range. The main goal of this section is the derivation of the explicit form of the right-hand sides of (3.1,5,36,37). The simplicity of the model for the medium allows one to see all the steps of the calculation of the quadratic terms of the macroscopic magnetic moment. However, the model is also realistic and therefore can be interesting in considering applied problems. Moreover, generalizations are possible, if one replaces the nuclear magnetic moment by any appropriate magnetic or electric moment, and the magnetic field by an electric field. Suppose further that the nucleus-field interaction is caused by the nuclear magnetic moment with the operator $\hat{\mu} = \gamma I$ and the interaction of a nucleus with the electron environment is described by the nuclear quadrupole moment and the tensor of the field gradient of the electronic shell. A constant field H_0 is also possible. The general scheme and the details of the calculation will be checked on the case where $H_0 = 0$, the formulas for the calculation with $H_0 \neq 0$ are given in Appendix 4. The Hamiltonian for a nucleus interacting with an electron cloud is well-known:

$$\mathcal{H}_Q = \frac{eQq_{zz}}{4J(2J-1)} \left[3I_3^2 + \frac{\eta}{2}(I_+^2 - I_-^2) - I^2 \right]$$
$$= AI_3^2 + B(I_+^2 - I_-^2) + CI^2, \tag{3.38}$$

where J is the nuclear spin, I_3, $I_x = (1/2)(I_+ - I_-)$, $I_y = (1/2i)(I_+ - I_-)$ are the projections of the spin operators, eQ is a scalar nuclear quadrupole moment, $\eta = |(q_{xx} - q_{yy})/q_{zz}|$ is the asymmetry parameter of the tensor gradient of the electric field q_{ik}. The evolution of the quantum system in an external alternating magnetic field is determined by the interaction operator (3.4). The last term CI^2 in (3.38) commutes with (3.4) and does not influence the calculations. We shall omit it.

Let us derive the eigenvector basis of the operator $\mathcal{H}_Q$ and express the interaction operator V in this basis. The operators I_k are usually determined in a normalized basis of eigenvectors of

$$I_3|m\rangle = m|m\rangle, \quad \alpha_m \equiv \sqrt{(J+m)(J-m+1)},$$
$$I_+|m-1\rangle = \alpha_m|m\rangle,$$
$$I_-|m\rangle = \alpha_m|m-1\rangle. \tag{3.39}$$

Obviously $(I_+^2 - I_-^2)|m\rangle = \alpha_{m+1}\alpha_{m+2}|m+2\rangle - \alpha_m\alpha_{m-1}|m-2\rangle$ or goes to zero if $m+2 > J$, $m-2 < -J$ in the first and second terms correspondingly. Substituting (3.38,39) and this result into the spectral equation

$$\mathcal{H}_Q|\lambda\rangle = E_\lambda|\lambda\rangle,$$

where $|\lambda\rangle = \sum_m c_m|m\rangle$, we get eigenvectors and eigenvalues of $\mathcal{H}_Q$. When $J = 1$ the vector $|0\rangle$ is the eigenvector of $\mathcal{H}_Q \sim E = 0$ and

$$|\pm\rangle = (|1\rangle \pm |-1\rangle)/\sqrt{2}, \quad |3\rangle = |0\rangle \tag{3.40}$$

are eigenvectors for the energy values $E_{1,2} = A \pm 2B$. Matrices $\langle\lambda|I_k|\lambda'\rangle$ in the basis (3.40) have the form

$$I_1 = \begin{pmatrix} 0 & 0 & 1 \\ 0 & 0 & 0 \\ 1 & 0 & 0 \end{pmatrix}, \quad I_2 = \begin{pmatrix} 0 & 0 & 0 \\ 0 & 0 & -i \\ 0 & i & 0 \end{pmatrix}, \quad I_3 = \begin{pmatrix} 0 & 1 & 0 \\ 1 & 0 & 0 \\ 0 & 0 & 0 \end{pmatrix}. \tag{3.41}$$

The interaction operator and the equilibrium density matrix are

$$V = -\gamma \begin{pmatrix} 0 & H_z & H_x \\ H_z & 0 & -iH_y \\ H_x & iH_y & 0 \end{pmatrix}, \quad \varrho^{(0)} = \varrho(0) = \text{diag}(\varrho_1, \varrho_2, 1).$$

Relaxation processes are included by the last term in (3.3). The relaxation processes cause the evolution of the diagonal matrix elements to equilibrium elements:

$$\dot{\varrho}_{ss} = [V, \varrho]_{ss}/i\hbar + (\varrho_{ss}^{(0)} - \varrho_{ss})/T_s ,$$

and nondiagonal terms go to zero. If we denote $\varrho' = \varrho - \varrho^{(0)}$ then the transition of the interaction representation is given by the substitution

$$\varrho'_{sk} = e^{-i(\omega_{sk} - i/T_{sk})t} \varrho^b_{sk}$$

where $\omega_{sk} = (E_s - E_k)/\hbar$. For the matrix ϱ^b

$$\dot{\varrho}^b_{sk} = \frac{1}{i\hbar} \exp i(\omega_{sk} - i/T_{sk})t[V, \varrho^{(0)} + \varrho']_{sk} \tag{3.42}$$

holds. The first order perturbation theory expression can be obtained if one neglects ϱ' in the right-hand side. Therefore, if the field is switched on at $t = 0$ we have

$$\varrho^{b(1)}_{sk} = \int_0^t \exp i(\omega_{sk} - i/T_{sk})t_1 [V, \varrho^0]_{sk} \, dt_1/i\hbar .$$

Calculating the commutator and denoting $\varrho_2 - \varrho_1 = \Delta\varrho$,

$$[V, \varrho^0] = -\gamma \begin{pmatrix} 0 & \Delta\varrho H_z & (1-\varrho_1)H_x \\ -\Delta\varrho H_z & 0 & -i(1-\varrho_2)H_y \\ -(1-\varrho_1)H_x & -i(1-\varrho_2)H_y & 0 \end{pmatrix} ,$$

we get the expression for the matrix elements $\varrho^{(1)}$ and the average value of $\mu^{(1)}$. For instance,

$$\mu^{(1)}_z = Sp\mu_z \varrho^{(1)} = 2\gamma^2 \Delta\varrho \int_0^t \exp[(t_1-t)/T_{12}] \sin \omega_{21}(t_1-t) H_z(t_1) dt_1 ,$$

which, after macroscopic averaging, is an example of a constitutive equation.

The next step in the derivation of the model equation is the concretization of the form for the components H_i. In the region of strong dispersion, as in the case of waveguide propagation, the behavior of the wave disturbance obeys the equations for the amplitudes of wave trains. If we choose H_i as in Sect. 3.1 (and Sect. 2.7) then taking into account only monodirectional waves [see notation of (3.39)] we have

$$H_3 = H_z = Z e^{i(\omega_3 t - k_3 x)} + \text{c.c.} ,$$
$$H_2 = H_y = Y e_2 + \text{c.c.} , \quad H_1 = H_x = X e_1 + \text{c.c.} . \tag{3.43}$$

We assume that the basic mode functions enter into X, Y, Z which contain small parameters describing slow variations of the amplitudes in time and coordinate x as in (3.32). Nonresonant terms are omitted in (3.43) despite their presence in the field components, as follows from the mode structure (3.10–12).

Let us denote $\Gamma_{sk} \equiv 1/T_{sk}$ and

$$\varrho^{(1)}_{12} = \frac{\Delta\varrho\gamma}{2\hbar} e^{-(i\omega_{12} + \Gamma_{12})t} \int_0^t e^{(i\omega_{12} + \Gamma_{12})t}(Ze_3 - \text{c.c.}) dt_1 . \tag{3.44}$$

Analogous formulas hold for $\varrho_{13}^{(1)}$, $\varrho_{23}^{(1)}$. One of the approximate integration results is

$$\varrho_{23}^{(1)} = Y \frac{\gamma(1-\varrho_2)}{2i\hbar} \left\{ \frac{e_2 - \exp[-i(\omega_{23}t + k_2 x) - \Gamma_{23} t]}{i(\omega_{23} + \omega_2) + \Gamma_{23}} \right.$$

$$\left. - \frac{e_2 - \exp[-i(\omega_{23}t - k_2 x) - \Gamma_{23} t]}{i(\omega_{23} - \omega_2) + \Gamma_{23}} \right\}; \qquad (3.45)$$

Such formulas are needed for $\varrho^{(2)}$ calculation. Expression (3.45) contains typical relaxation contributions. The values of Γ_{ik} in these terms allow one to estimate their significance in the description of the evolution. The commutator $[V, \varrho^{(1)}]$ is calculated and the equation for $\dot{\varrho}^{(2)}$ is integrated with (3.44). Then we sum up diagonal terms of $\gamma I_z \varrho^{(2)}$. We get

$$\mu_z^{(2)} = Sp\gamma I_z \varrho^{(2)} = \gamma(\varrho_{21}^{(2)} + \varrho_{12}^{(2)})$$

$$= -\frac{\gamma^2}{i\hbar} \left\{ e^{-(i\omega_{12}+\Gamma_{12})t} \int_0^t e^{(i\omega_{12}-\Gamma_{12})t_2} \left[H_x(t_2)\overline{\varrho_{23}^{(1)}}(t_2) \right. \right.$$

$$\left. - i\varrho_{13}^{(1)}(t_2) H_y(t_2) \right] dt_2 + e^{(i\omega_{12}-\Gamma_{12})t} \int_0^t e^{(i\omega_{12}+\Gamma_{12})t_2}$$

$$\times \left[H_x(t_2)\varrho_{23}^{(1)}(t_2) - i\overline{\varrho_{13}^{(1)}} H_y(t_2) \right] dt_2 \right\}. \qquad (3.46)$$

Similar expressions can be written for $\mu_y^{(2)}$, $\mu_x^{(2)}$. We can see that $\mu_z^{(2)}$ is expressed by additional projections of the magnetic field because $\varrho_{23}^{(1)} \sim Y$ by (3.45) and similarly $\varrho_{13}^{(1)} \sim X$. Further, it is necessary to include H_x, H_y, $\varrho_{ik}^{(1)}$ from (3.43–45) in (3.46) and then integrate approximately, by removing X, Y from the integrand. This is possible because of the slow variation of these functions compared with the exponents. The general expression is directly obtained but it is quite complicated.

As follows from the derived formulas, e.g., (3.45), the intermode interaction effect is resonant in the usual sense of the word: when the frequency of the external field ω coincides with the transition frequency of the medium, the quantum system, the nonlinear constant reaches its maximum. In our case the electromagnetic wave interacts with a three-level system. Therefore, there are three resonant frequencies ω_{23}, ω_{13}, ω_{12} and

$$\omega_{13} = \omega_{23} + \omega_{12} \qquad (3.47)$$

due to ω_{sk} determination by E_i. The equality (3.47) is the three-wave resonance condition if ω_i coincides with the transition frequencies $\omega_1 = \omega_{13}$, $\omega_2 = \omega_{23}$, $\omega_3 = \omega_{12}$ [3.6]. Namely, when this condition is fulfilled, oscillating exponents are cancelled and the interaction effect is stored in time. The integration result for the resonant term in (3.46) can be written as

$$\frac{\gamma XY \Delta \varrho \left(e^{-\Gamma_{12} t} - 1 \right) e^{i(k_2 - k_1)x}}{4\hbar i \Gamma_{12}(\omega_{13} + \omega_{23})}$$

if $\omega_{23} + \omega_2 = \omega_{13} + \omega_1$ and $\omega_{23} + \omega_{13} \gg \Gamma_{23}, \Gamma_{13}$. Analogously, if all the resonant conditions are valid then

$$\varrho_{13}^{(2)} = e^{-i\omega_{13}t} \frac{\gamma ZY}{4i\hbar} \left\{ \left[\frac{1 - e^{-\Gamma_{13}t}}{i\Gamma_{23}\Gamma_{13}} - \frac{e^{-\Gamma_{23}t} - e^{-\Gamma_{13}t}}{i\Gamma_{23}(\Gamma_{13} - \Gamma_{23})} \right] (1 - \varrho_2) \right.$$

$$\left. + i\Delta\varrho \frac{1 - e^{-\Gamma_{13}t}}{i\Gamma_{12}\Gamma_{13}} \right\} e^{i(\varphi_2 - \varphi_3)} .$$

If, moreover, $t \gg T_{13}, T_{23}$, then

$$\varrho_{12}^{(2)} = \frac{\gamma^2 X \bar{Y}}{4i\hbar^2 \Gamma_{12}} \left(\frac{1 - \varrho_2}{\Gamma_{23}} - \frac{1 - \varrho_1}{\Gamma_{13}} \right) \exp i[\omega_{12}t + (k_1 - k_2)x] , \quad (3.48)$$

$$\varrho_{13}^{(2)} = \frac{\gamma^2 YZ}{i4\hbar^2 \Gamma_{13}} \left(\frac{\Delta\varrho}{\Gamma_{12}} - \frac{1 - \varrho_2}{\Gamma_{23}} \right) \exp i[\omega_{13}t + (k_2 + k_3)x] , \quad (3.49)$$

$$\varrho_{23}^{(2)} = \frac{\gamma^2 \bar{Z} X}{i4\hbar^2 \Gamma_{23}} \left(\frac{1 - \varrho_1}{\Gamma_{13}} - \frac{\Delta\varrho}{\Gamma_{12}} \right) \exp i[\omega_{23}t + (k_1 + k_3)x] , \quad (3.50)$$

Let us denote relaxation factors as

$$P_x = [(\varrho_1 - \varrho_2)T_{12} + (1 - \varrho_2)T_{23}] T_{13} ,$$
$$P_y = [(\varrho_2 - \varrho_1)T_{12} + (1 - \varrho_1)T_{13}] T_{23} ,$$
$$P_z = -[(1 - \varrho_2)T_{23} + (1 - \varrho_1)T_{13}] T_{12} .$$

Substituting (3.48–50) and analogous formulas for $\mu_x^{(2)}$, $\mu_y^{(2)}$, we have

$$\mu_x^{(2)} = i\gamma^3 P_x ZY \exp -i(\omega_1 t - k_2 x - k_3 x)/\hbar^2 \text{c.c.} ,$$
$$\mu_y^{(2)} = i\gamma^3 P_y X \bar{Z} \exp -i(\omega_2 t - k_3 x + k_1 x)/\hbar^2 + \text{c.c.} ,$$
$$\mu_z^{(2)} = i\gamma^3 P_z X \bar{Y} \exp -i(\omega_3 t + k_1 x - k_2 x)/\hbar^2 + \text{c.c.} . \quad (3.51)$$

The last stage of the derivation is the substitution of $M^{(2)} = N\mu^{(2)}$ into (3.5) and of the expressions for the magnetic field components into the linear part followed by the subsequent projection onto the basis vectors of the modes using (3.32–37). The details of this procedure depend on the choice of the experimental method, for example, whether a transmission or double resonance spectrum is measured [3.2,7]. If the experiment is done in a rectangular metal waveguide then the method of Sects. 2.1,2 is convenient and results in equations of the same type as (3.36) with the substitutions in the nonlinear terms in which (3.51) is taken into account. (See also Appendix 4, where the possibilities of waveguide resonances are investigated.) Technically the simplest experiment is on dielectric waveguides. The peculiarities of the wave propagation in such waveguides are discussed in Sect. 3.4. NQR-NMR resonances are observed in the meter wavelength range, therefore, the transverse length of a metal waveguide is rather large if one or two modes are allowed. In a dielectric waveguide only one mode exists at any frequency of the generator. Pulse doubling may be observed due to a nonlinear effect in which a slow propagating mode is excited, at a given delay time, and the input pulse will be transformed into two output pulses. Theoretical and technical questions in these experiments still remain, but in any case, the specific mode content can influence only the "projecting coefficients of unit order in constants of the three-wave (n-wave) interaction". This coefficient is the scalar product of the mode basis functions. Namely, if $X \to XC_1$, $Y \to YC_2$, $Z \to ZC_3$, where C_i are the normalized mode basis functions, then the final form of the model system is [3.8]

$$X_t + v_1 X_x = i\beta_1 ZY \exp(-i\Delta kx),$$
$$Y_t + v_2 Y_x = i\beta_2 X\bar{Z} \exp(i\Delta kx),$$
$$Z_t + v_3 Z_x = i\beta_3 \bar{Y} X \exp(i\Delta kx), \qquad (3.52)$$

where $\Delta k = k_1 - k_2 - k_3$, $\beta_i = \varepsilon\gamma^3 \omega_i^3 P_i v_i N(C_i, C_k C_e)/2k_i\hbar^2 c^2$, i, k, l form a cyclic permutation. The nonlinear constants satisfy $\beta_{1,2} > 0$, $\beta_3 < 0$ and the effect of parametric amplifying occurs [3.2,7]. The expression of the operator V that is given in Appendix 4 (A4.8) shows what influences an increase in the complexity of the nonperturbed Hamiltonian can have. One can choose the structure of the wave trains so that the nonlinear resonance condition distinguishes a three-wave system. The spin 3/2 case and the calculation of the nonlinear constants were elaborated in the thesis of *V. Malaschenko* (Kaliningrad Univ. 1981).

The derivation of the three-wave waveguide mode interaction can be directly transformed to the case of electric interaction. It is enough to change the magnetic field and momentum vectors to electric field vectors. The sequence of steps is the same, only the designations and results of the commutations are changed. The pseudospin methodology is convenient for a few level systems [3.9]. This calculation technique can be used for media that consist of a large number of classical or quantum subsystems such as dislocations in solids and powders of dielectrics or ferromagnets. Nonlinear terms are written directly in adjoint representation of the Poisson bracket algebra of the dynamical variables [3.10].

The solution of Cauchy problems for systems such as (3.52) in the class of decreasing functions is developed in [3.11]. In that book multi-soliton solutions are given. The space synchronism condition is not necessary and for this version of the system (3.52) with $\Delta k \neq 0$ the solutions are given in [3.12]. In Sect. 2.7 it is shown that a three-wave system at $v_2 = v_3$, $\Delta k = 0$ may be transformed into the sine-Gordon equation (SG) [3.8,13]. With this transformation the soliton solutions can be written as well and any localized initial or boundary condition can be analyzed for soliton content [3.14–19]. If $v_1 = v_3$ the system (3.52) can be transformed to the Liouville equation by the substitution [3.8]

$$\psi_\xi = 2X\sqrt{-\beta_2\beta_3}, \quad e^\psi = |\beta_1\beta_2 Y|^2 + |\beta_1\beta_3 Z|^2,$$
$$\xi = (x - v_2 t)/(v_1 - v_2), \quad \tau = (x - v_1 t)/(v_1 - v_2).$$

The Liouville equation $\psi_{\xi\tau} = -e^\psi$ is described in [3.20].

The analysis of the solution parameters for (3.52) that are related to the SG soliton [3.21] when $v_2 = v_3 = u$, $v_1 = v$, $\zeta = t - x/v_c$

$$X = iT^{-1}(1 - u/v_c)(\beta_2\beta_3)^{-1/2}\text{sech}(\zeta/T) ,$$
$$Y = -Y_0 \tanh(\zeta/T) ,$$
$$Z = Y_0\beta_3^{1/2}\beta_2^{-1/2}\text{sech}(\zeta/T) \tag{3.53}$$

guides the choice of the experimental conditions for the detection of mode interaction. Here v_c is the soliton velocity. The structure of the solution of (3.53) is such that a change of the polarity of a radiofrequency voltage in the Y component during time T, as in (3.53), and the simultaneous input of a pulse of known width in the field Z leads to the appearance of a pulse in the X-component with a known amplitude. By finding the soliton contribution and by measuring the amplitude, one can determine the nonlinear constants β_i and therefore the relaxation times T_{ik}. The parameters of the solution (3.53) are related among themselves by the equality

$$\beta_1\beta_3|Y_0|^2T^2 = (1 - v/v_c)(u/v_c - 1) . \tag{3.54}$$

It can be seen that the velocity of the soliton v_c may vary up to a few orders of magnitude at realistic amplitudes Y_0 and constants β_i [3.8]. If the ratio $\omega_{ik}/\Gamma_{ks} \sim 10^3 - 10^4$, and the temperature $T \sim 1 - 300K$ then $\beta_i \sim 10^{-2} - 10^{-6}$. When the values of the group velocities are close $(v - u)/u \sim 10^{-2}$ then the velocity of soliton propagation is determined by the dimensionless constant $s = (\beta_1\beta_3)^{1/2}\tau_c Y_0$. If the pulse duration is $\tau_c \sim 10^{-3}s$ and $Y_0, Z_0 \sim 10^2 - 10^3 A/m$ then $s = 10^{-2}$, $v_c/u \approx 10^{-2}$ and the soliton velocity is much less than the group velocity of the linear wave. The amplitude of the generated wave is about $X_0 = 10^3 A/m$.

3.4 Nonlinear Mode Dispersion in Dielectric Waveguides

Let us consider electromagnetic wave propagation inside a dielectric guide [3.2,4]. Waveguides such as dielectric fibers, are widely applied in the range of optical wavelengths [3.22,23]. The problem of nonlinear pulse transmission in dispersive dielectric fibers is of great interest both from physical and technical points of view [3.24–26] as well as from the viewpoint of mathematical physics [3.27–29]. New modifications of the NS equation (3.31) and the systems (3.36,37) with different linear and nonlinear dispersion terms have been derived and investigated [3.30,31]. The dispersion terms are obtained by means of perturbation theory and are of a mixed type, as in the case of the interacting Rossby–Poincaré waves (2.49–50). It should be noted that like in the CKdV situation the modification of the evolution equation retains the basic soliton properties. For example, an initial condition decays into individual stable solitons. This property gives the possibility of generation and transmission over long distances of optical pulses of short duration, up to 16 femtoseconds [3.22,32,33]. Waveguide propagation of electromagnetic waves in dielectrics, with its possibilities of nonlinearity and dispersion control, provides opportunities for investigation of wavefront inversion and real time holography [3.34,35].

The geometry of the example to be considered in this section is the simplest possible but the generalizations are obvious. We discuss a strongly inhomogeneous medium (Chap. 5) and shall show the operations that lead to the model equation that contains an integral (pseudodifferential) dispersion operator. We note once more the success of mathematical methods of integration of such equations [3.36–38].

Let a dielectric plate with the dielectric constant ε span the interval $z \in [-h, h]$. We assume that outside this interval the constant ε is equal to 1. Now x is the axis of propagation and the field does not depend on y. In classical theory an electric field with frequency ω_0 and the longitudinal wave vector $\mathbf{k}_o$ is described by the Hertz vector $\mathbf{\Pi}$ that is directed along x [3.1]. The appropriate projection is denoted as $\Pi_x \equiv \Pi$. Then inside the dielectric layer

$$\Delta \Pi + \omega_0^2 \varepsilon \Pi / c^2 = 0 , \qquad (3.55)$$

is valid and outside, the analogous equation with $\varepsilon = 1$. The mode representation that appears when the variables are shared is characterzed by the dependence on z that is determined by (3.55) and the boundary conditions. Using the relationship between the electric field components and the Hertz vector and (3.55), one can conclude that

$$S_\varepsilon \Pi \big|_{h-0} \equiv (k_0^2 + \varepsilon \omega_0^2/c^2) \Pi \big|_{h-0} = (k_0^2 + \omega_0^2/c^2) \Pi \big|_{h+0} \equiv S_0 \Pi \big|_{h+0} ,$$
$$\varepsilon \Pi_z \big|_{h-0} = \Pi_z \big|_{h+0} . \qquad (3.56)$$

For the Fourier components (plane waves) at $|z| > h$, $\Pi = \exp(-p|z| + ik_0 x)$, we have

$$k_0^2 - \omega_0^2/c^2 = p^2, \quad \mp \Pi_z \big|_{\pm h \mp 0} = S_\varepsilon p \Pi / (\varepsilon S_0) \big|_{\pm h \mp 0} . \qquad (3.57)$$

The dispersion relation

$$\varepsilon \omega_0^2 / c^2 = \alpha^2 + k_0^2 \qquad (3.58)$$

relates the wave number α with k_0 and ω_0. Boundary conditions give

$$\varepsilon^{-1} \alpha h \tan(\alpha h) = ph . \qquad (3.59)$$

The corollary of (3.57) and (3.58) is

$$(\alpha h)^2 + (ph)^2 = \omega_0^2 h^2 (\varepsilon - 1)/c^2 . \qquad (3.60)$$

Equations (3.59,60) in the region $ph > 0$ have compatible solutions at the points of intersection of tangent curves in the intervals $n\pi < \alpha h \leq n\pi + \pi/2$, $n = 0, 1, \ldots$ with the circumferences (3.60). At a given ω_0 the number of crossing points is finite and determines the mode number with the basis functions $Z_\alpha(z)$ that satisfy $Z_\alpha'' = -\alpha^2 Z_\alpha$, $\int Z_\alpha^2 dz = h$.

The basis functions that are constructed in the linear approximation and the boundary conditions (3.57) are used for generalization to the nonlinear case. The finiteness of the spectral width of the boundary (along x) and of the initial conditions leads to wave packet dispersion. The linear dispersion could be described by a superposition of Fourier components. For the solution of the nonlinear dispersion problem the model evolution equations are used because they are simple to analyze. Internal waves in inhomogeneous water layers (thermoclyne or pycnoclyne) were successfully described with nonlocal analogues of KdV equations, equations of *Benjamin–Ono* [3.39], *Joseph* [3.36] and their two-dimensional generalizations [3.40–42].

The most preferable region for the suggested technique in the example to be discussed is the region in which $k_0^2 \ll \alpha^2$ and approximately $\omega_0 =$ const. The radius of the circumference in (3.60) varies weakly and therefore the value of the spectral parameter α is almost constant. This means that the Helmholtz equation can be used in the region of $|z| > h$. We note that the generalization taking into account the variation in α is possible if the Klein–Fock equation is used. Thus the variation of the spectral parameter composition requires a general solution for a "quasi-waveguide" at $|z| \geq h$ and therefore it influences, through the boundaries, the nonlinear dispersion characteristics inside $z < h$. Then we should find an approximate solution in the form

$$\Pi = \sigma Z_\alpha(z) A(\beta x, \beta t) \exp i(k_0 x - \omega_0 t) + \text{c.c.} + \sigma \beta^2 \tilde{\Pi} \ . \tag{3.61}$$

Supposing that the "linear" excitation conditions of a single mode contribution are met and a deviation from the single-mode state $\tilde{\Pi}$ is stipulated by weak nonlinearity and interaction with the environment of the dielectric layer. We choose an initial state such that the dispersion parameter β is small and $\sigma \sim \beta^2 \ll 1$. The factors mentioned above allow one to get, for a given dielectric type, an expression for the nonlinear contribution $\sigma N(\Pi)$ in the wave equation

$$\Delta \Pi - \varepsilon \Pi_{tt}/c^2 = \sigma^2 N(\Pi) \ . \tag{3.62}$$

After substitution of (3.61) into (3.62) we have, due to the dispersion relation (3.58)

$$Z_\alpha \left(ik_0 A_x - \frac{i\varepsilon\omega_0 A_t}{c^2} + \frac{\beta(A_{xx} - \varepsilon A_{tt})}{c^2} \right) e^{i(k_0 x - \omega_0 t)}$$
$$+ \text{c.c.} + \beta \tilde{\Pi}_{zz} = \beta^{-1} \sigma N(\Pi) \ . \tag{3.63}$$

Because the parameters β and σ are small, the approximation $A_{xx} = \omega_0^2 \varepsilon^2 A_{tt} c^4 k_0^2$ holds. The projection to the mode subspace is given by a scalar multiplication of Z_α. In multiplying (3.63) by Z_α and integrating over z between $-h$ and h, we use the normalization condition of Z_α and linear independence of the exponents $\exp[-i(k_0 x - \omega_0 t)]$. We get

$$ik_0 A_x - i\varepsilon\omega_0 A_t/c^2 + \beta\varepsilon\alpha^2 A_{tt}/c^2 k_0^2$$
$$+ e^{-i(k_0 x - \omega_0 t)} \left[Z_\alpha(z)(\tilde{\Pi}_{1z} \pm pS_\varepsilon \tilde{\Pi}_1/\varepsilon S_0) \right]_{-h}^{h} = \beta^{-1}\sigma N_1 \ . \tag{3.64}$$

The dispersion relation is used too. Here N_1 is the projection of the nonlinear contribution on the subspace $Z_\alpha \exp[i(k_0 x - \omega_0 t)]$ that is usually generated by a cubic nonlinearity [3.2,14]. The $\tilde{\Pi}_1$ is a correction that is connected with the "layer interaction". The last term in the left side of (3.64) appears after integration by parts. Let us calculate this addend using iterations in the parameter β in the boundary conditions.

To this end we solve the Dirichlet problem for the Helmholtz equation with $\varepsilon = 1$ in the halfplane $z > 0$. The surface Green's function is

$$G_s = \frac{i\omega_0}{2c} H_1^{(1)} \left[\frac{\omega_0}{c}\sqrt{z^2 + (x-\eta^2)}\right] \frac{z}{\sqrt{z^2 + (x-\eta)^2}}. \qquad (3.65)$$

Here $H_1^{(1)}$ is a Hankel function of the first kind. After shift and reflection of the coordinate z in (3.65) we get the Green's function for both regions $z > h$, $z < -h$. For the upper boundary

$$\Pi_{1z}^+ \Big|_{z=h} = \frac{i\omega_0}{2c} \int_{-\infty}^{\infty} H_1^{(1)} \left(\frac{\omega_0}{c}|x-\eta|\right) \frac{b_+ d\eta}{|x-\eta|}, \qquad (3.66)$$

where $b_+ = Z_\alpha(h)A(\beta\eta, \beta t)\exp[i(k_0\eta - \omega_0 t)]$. The derivative at the lower boundary is analogous. Substituting the results in (3.64) we find

$$iA_x - \frac{i\varepsilon\omega_0 A_t}{k_0 c^2} + \frac{\beta\varepsilon\alpha^2 A_{tt}}{c^2 k_0^3}$$

$$+ \frac{i\omega_0}{2ck_0}\left[Z_\alpha^2(h) + Z_\alpha^2(-h)\right]\left[\int_{-\infty}^{\infty} H_1^{(1)}\left(\frac{\omega_0}{c}|x-\eta|\right)\right.$$

$$\left. \times \frac{A(\eta,t)e^{ik_0(\eta-x)}d\eta}{|x-\eta|} + \frac{pS_\varepsilon}{\varepsilon S_0}A\right] = \frac{\sigma N_1}{\beta k_0}. \qquad (3.67)$$

In this model equation, which generalizes the nonlinear Schrödinger equation, an x-translation invariant pseudodifferential operator contributes in the dispersion term.

The theory of waveguide propagation shows the important possibilities of detecting nonlinear effects. Interaction leads to excitation of modes that are initially absent. The theory of waveguide mode interaction introduces the systems of equations [3.4,43,44] that are similar to well-known n-wave systems [3.11]. As is clear from the mentioned example, the dispersion laws and nonlinear operators of such systems may depend on the boundary conditions.

We shall not write the entire formal procedure, as the considered examples describe all the principal features of the approach. We only note in addition that the method can be generalized to the case of the coefficients of a coordinate-dependent noncommuting operator in the fundamental evolution equation if the parameter limits allow one to separate the individual wave contributions. The projection operator for constant coefficients of the equation $L\psi = \lambda\psi$ is determined by $\psi^{(i)} = \{\alpha_{ik}\}$ which is an eigenvector. The projection operator onto the subspace of $\psi^{(i)}$ is then

$$P_{ik}^{(m)} = \alpha_{mi}P_{mk} \qquad (3.68)$$

where $P^T_{km}\alpha_{mi} = \delta_{ki}$, i.e., P_{mk} is the reciprocal matrix transposed to α. For example, for $n = 2$, in the convenient norm

$$\psi^{(1)} = \begin{pmatrix} 1 \\ b \end{pmatrix}, \quad \psi^{(2)} = \begin{pmatrix} 1 \\ d \end{pmatrix}, \quad P^{(1)} = \begin{pmatrix} d & -c \\ bd & -bc \end{pmatrix} \frac{1}{d-b}.$$

For noncommuting elements of L it is necessary to solve the equations for $P^{(m)}_{ik}$ sequentially. Namely, if $\psi_1^{(i)} = 1$ and $\psi_2^{(i)} = a_i$

$$P^{(i)} = \begin{pmatrix} \alpha_i & \beta_i \\ a_i\alpha_i & a_i\beta_i \end{pmatrix}$$

$\alpha_1 = 1 - (a_1 - a_2)^{-1}a_1$, $\alpha_2 = 1 - (a_2 - a_1)^{-1}a_2$, $\beta_1 = (a_1 - a_2)^{-1}$, $\beta_2 = -\beta_1$.

In conclusion we add that a case exists, of stratified dielectric slabs, for which it is possible for a soliton to penetrate through the boundary slab [3.45,46].

4. Nonlinear Waves in Stratified Plasma

The main aim of this chapter is the derivation and further analysis of the evolution equation for describing the Langmuir and ion-acoustic waves using two variants of the theory. The first variant is deterministic (Sect. 4.1,2) and the second one is based on the averaging procedure that is arranged for the turbulent plasma. Special attention is paid to the perturbation procedure technique for the kinetic description of plasma waves (Sect. 4.3). The straight algebraic treatment allows one to find explicit (soliton-like) solutions (Sect. 4.4) for the one-dimensional case of the Langmuir turbulence equations taking into account the radiation and absorption of ion-acoustic waves.

4.1 Wave Modes in Stratified Plasma

4.1.1 Description of Plasma Waves

Let us consider the simplest type of propagation of plasma waves without treating all the entire variety of plasma oscillations [4.1-6]. The content of this chapter is based on a considerable extent on the textbook by *Silin* [4.4]. The theory of plasma waves is based on the electromagnetic field equations (3.6) that are completed by material equations in the form

$$D = \mathcal{E} + 4\pi \int_{-\infty}^{t} dt_1 j(r, t_1) \tag{4.1}$$

where j is a current density that is a function of an electric field vector $\mathcal{E}$. It is assumed that dependence on the magnetic field is excluded by a suitable equation for B. The form of the material equation is determined by the plasma state and is generally nonlocal due to spatial dispersion. The macroscopic state of the plasma is described by a density matrix that is a function of phase space $\{r_a, p_a\}$ and spin $\{\sigma_a\}$ coordinates (Wigner representation [4.4]). Such a representation is convenient for a transition to a classical description. We restrict ourselves, however, to a classical description so that we may use the ion distribution functions and calculate densities of charge and current by integrating these functions over momenta and summing over the charge types. The closed system of equations for the electric field components that are in the distribution function has the form

$$\text{div}\,\mathcal{E} = 4\pi \sum_a e_a \int dp_a f_a , \tag{4.2}$$

$$\frac{1}{c}\frac{\partial \mathcal{E}}{\partial t} = \text{rot } \boldsymbol{B} - \frac{4\pi}{c}\sum_a e_a \int d\boldsymbol{p}_a f_a \boldsymbol{v}_a , \qquad (4.3)$$

$$\frac{1}{c}\frac{\partial \boldsymbol{B}}{\partial t} = -\text{rot } \mathcal{E} , \quad \text{div } \boldsymbol{B} = 0 .$$

When the Knudsen number Kn (the relation of the mean free path of the particles to the wavelength) is large, the distribution function satisfies the Vlasov equation [4.4,5] in which interaction of particles is accounted for by the self-correlated Lorentz force field $\boldsymbol{F}_a = e_a(\mathcal{E} + [\boldsymbol{v}_a \times \boldsymbol{B}]/c)$,

$$\frac{\partial f_a}{\partial t} + \boldsymbol{v}_a \frac{\partial f_a}{\partial \boldsymbol{r}_a} + \boldsymbol{F}_a \frac{\partial f_a}{\partial \boldsymbol{p}_a} = 0 \qquad (4.4)$$

$\boldsymbol{v}_a = \boldsymbol{p}_a/m_a$ is the velocity, e_a is the charge of the plasma particle of type a [4.7]. In the framework of the Vlasov theory the Debye radius, an important concept in plasma physics, is introduced

$$r_\text{D} = \left(\sum_a 1/r_{\text{D}a}^2\right)^{-1/2} = \left(\sum_a \frac{4\pi e_a^2 n_a}{\kappa_\text{B} T_a}\right)^{-1/2} . \qquad (4.5)$$

in (4.5) $r_{\text{D}a}$ denotes the screening radius for a component, n_a is the concentration, T_a temperature, κ_B the Boltzmann constant. On the basis of self-correlated field theory it is possible to build an elementary theory of plasma oscillations. It can be demonstrated that (4.4) is compatible with (4.2,3).

An important parameter for the analysis of media dispersion properties is the penetrability tensor. In the linear theory it is determined by the equality

$$D_i(\boldsymbol{r}, t) = \int_{-\infty}^{t} dt_1 \int d\boldsymbol{r}_1 \varepsilon_{ij}(\boldsymbol{r}, \boldsymbol{r}_1, t, t_1) \mathcal{E}_j(\boldsymbol{r}_1, t_1) . \qquad (4.6)$$

As follows from (4.3), knowledge of the distribution function dependence on $\mathcal{E}_j$ allows one to find the tensor ε_{ij}. The Fourier transform of the field equation without sources gives

$$k_i \varepsilon_{ij}(\omega, \boldsymbol{k}) \mathcal{E}_j = 0, \quad [\boldsymbol{k} \times \mathcal{E}] = \omega \boldsymbol{B}/c ,$$
$$[\boldsymbol{k} \times \boldsymbol{B}]_i = \omega \varepsilon_{ij}(\omega, \boldsymbol{k}) \mathcal{E}_j/c, \quad (\boldsymbol{k}, \boldsymbol{B}) = 0 . \qquad (4.7)$$

The spatial dispersion leads to a dependency ε on $\boldsymbol{k}$ and, as corollary to the tensor coupling equations (4.6), between components of fields, even in the isotropic case. In the coordinate system oriented along the wave propagation axis the tensor ε_{ij} has two independent components ε^l and ε^tr and

$$\varepsilon_{ij} = \left(\delta_{ij} - \frac{k_i k_j}{k^2}\right)\varepsilon^\text{tr} + \frac{k_i k_j}{k^2}\varepsilon^l . \qquad (4.8)$$

The substitution of (4.8) into (4.7) in this coordinate system leads to solvability condition of the system (4.7) that consists of two equations

$$\varepsilon^l(\omega, k) = 0, \quad k^2 - \omega^2 \varepsilon^\text{tr}(\omega, k)/c^2 = 0 .$$

The roots of these equations determine the dispersion branches of plasma waves (Sects. 2.1,2). As follows from dispersion equations or directly from the system (4.2–4), in the linear theory transverse and longitudinal waves may be considered as independent. For longitudinal waves the closed system is restricted by (4.2) and (4.4) (under the condition of nonexcited transverse components of $\mathcal{E}'$ at the initial time). Indeed, the projection on the direction of (4.3) is the corollary of (4.2) and (4.4) as the integration over momenta nullifies the terms with a magnetic field in the Lorentz force.

In homogeneous collisionless plasma in the absence of an external field the density matrix of nonperturbed state f^0 can be independent of the coordinates. Wave disturbances of plasma are oscillations that are related to the motion of electrons and ions. Indeed, a displacement of an electron from the equilibrium state leads to the appearance of a "hole" – a positively charged region. Between charges there appears an electric field that generates an attractive restoring force which influences both positive and negative charges. We first consider the classical example of a wave in a cold plasma neglecting the ion motion. Let us write the kinetic equation (4.4) for electrons with a linearized Vlasov term in $\mathcal{E}$ and the distribution function χ. This corresponds to the first iteration in the equation for the density matrix in Sect. 3.3,

$$\chi_t + (\boldsymbol{v}, \nabla)\chi = -e(\mathcal{E}', \partial f^0/\partial \boldsymbol{p}) \equiv F . \tag{4.9}$$

The oscillations of the electric charge density determine $\mathcal{E}'$

$$\operatorname{div} \mathcal{E}' = 4\pi e \int d\boldsymbol{p}\chi . \tag{4.10}$$

Supposing that the electromagnetic field represents a wave packet propagating along the x axis we introduce for the longitudinal component

$$\mathcal{E}'_x \equiv \mathcal{E}' = A(\beta x, \beta t) e^{i(kx-\omega t)} + \text{c.c.} . \tag{4.11}$$

Without loss of generality the x axis is chosen such that its direction coincides with the direction of wave propagation. Equation (4.9) is integrated by the method of characteristics

$$\chi = \int_0^t F(x - p(t-\tau)/m, \tau) d\tau , \tag{4.12}$$

where m is the electron mass and p is the x-component of its momentum. Substituting (4.11) into (4.12) we get in the zeroth order (in the small parameter) approximation

$$\chi = e f_p^0 A \exp[i(kx - \omega t)]/i(\omega - pk/m) + \text{c.c.} . \tag{4.13}$$

Now both the current density j and the charge density can be found. Let us put (4.13) into (4.10) and equate coefficients of linearly independent exponents:

$$ikA = 4\pi e^2 A \int d\boldsymbol{p} f_p^0/i(\omega - pk/m) . \tag{4.14}$$

Under the conditions of cold plasma we expand the integrand in a power series up to $(kp/m\omega)^3$

$$1/(1 - pk/m\omega) = 1 + pk/m\omega + (pk/m\omega)^2 + (pk/m\omega)^3 + \ldots . \quad (4.15)$$

If the distribution function f^0 is even in p and decreases at infinity quickly enough then only odd powers from (4.15) contribute to the integral. Due to the normalization condition of the unperturbed distribution function

$$\int d\mathbf{p}\, f^0 = n , \quad (4.16)$$

the integrals of powers of the momentum component p are

$$\int p\, f_p^0\, dp\, dp_y\, dp_z = -n , \quad \int p^3\, f_p^0\, dp\, dp_y\, dp_z = -3n\langle p^2\rangle ,$$

where the bracket $\langle \ldots \rangle$ denotes averaging over the momentum. Therefore

$$-ik = 4\pi e^2 k n [1 + k^2\langle p^2\rangle/m^2\omega^2]/im\omega^2$$

and the dispersion relation follows. For the Maxwell distribution f^0 it has the form

$$\omega^2 = \omega_{L_e}^2 + 3\kappa_B T_e k^2/m \quad (4.17)$$

where $\omega_{L_e} = 2e\sqrt{\pi n/m}$ is the Langmuir frequency of the electrons and T_e is the electron temperature.

If $\varrho = \varrho_0 \exp(-i\omega t)$, $\sqrt{\langle p^2\rangle}\, k/m\omega \ll 1$ and the same condition is valid for ions, then (4.13) with ion terms gives

$$\chi = \sum_a e_a \frac{f_{p_{ax}}^{0a}}{i(\omega - p_{ax}k/m_a)} A e^{i(kx-\omega t)} + \text{c.c.} . \quad (4.18)$$

Because for longitudinal currents $\mathbf{j} = \sigma \mathcal{E}$

$$\text{div}\, \sigma \mathcal{E} = \sigma\, \text{div}\, \mathcal{E} = \sigma 4\pi \varrho ,$$

due to $\varrho_t = -\text{div}\, \mathbf{j}$, $-i\omega + 4\pi\sigma = 0$. Since $m_i \gg m_e$,

$$\omega^2 = \sum_a \frac{4\pi e_a^2 n_a}{m_a} \approx \frac{4\pi e^2 n}{m_e} \equiv \omega_{L_e}^2 .$$

Approximate equality occurs if there is no great difference between the concentrations n_a. For example, this is true for a single-ion quasineutral plasma. Obviously the Langmuir frequency is basic for longitudinal plasma oscillations in the long wavelength range.

For real ω the integral in (4.14) is singular. Therefore it is natural to assume that the frequency in (4.11) is a complex number and the wave amplitude decreases in time. A small imaginary part of the frequency may be found by the Sokhotsky formula

$$\int \frac{\varphi dp}{m\omega/k - p} = \oint \frac{\varphi dp}{m\omega/k - p} - i\pi\varphi\,; \quad (p = m\omega/k)\,.$$

For the Maxwell distribution, at $\omega = \omega' + i\omega''$ one gets for imaginary part:

$$\omega'' = -\sqrt{\pi/8}\frac{\omega_{L_e}}{(kr_{D_e})^2}\exp\left\{-\frac{(\omega')^2}{2k^2 v_{T_e}^2}\right\}\,. \tag{4.19}$$

The smallness of ω'' allows the wave to exist. The wave energy is transformed to plasma heating. As a result of (4.12) the distribution function is obviously changed and the mean quadratic velocity of the electrons (and temperature) increases. Because the main contribution to the integral is made by the region of momentum in which $p/m = \omega/k$ the energy transport occurs via the Cherenkov effect. Collisionless relaxation which is a result of the self-adjusting field is known as Landau damping [4.8].

Accounting for the movement means that in addition to (4.9) it is necessary to introduce an analogous equation for ions [see (4.4)] and conserve the sums in (4.2) and (4.3). The increase of the number of equations with time derivatives generally leads to new roots in the dispersion relation (Chaps. 1,2). The analysis of the $\omega_i(k)$ and of the corresponding dispersion relation is complicated by the different coordinate spaces used for the kinetic and electromagnetic equations. It is necessary to generalize the mathematics of splitting of waves and introduce projection operators onto subspaces with given $\omega_i(k)$. Here we use the traditional approach: we solve the distribution equations by perturbation theory and enter the results into the electrodynamic equations.

The existence of two dispersion branches for the single-ion plasma may be determined by analyzing the expression for the perturbed distribution function (4.18) generated by the wave motion. The dispersion equation can be derived from (4.2) or (4.10) after substitution of (4.18) for χ. For longitudinal waves we have

$$k^2 = \sum_a 4\pi e_a^2 m_a \int d\mathbf{p}_a \frac{f_{p_a}^{0a}}{p_a - m_a\omega/k}\,, \tag{4.20}$$

where p_a is the x-projection of the momentum of ion a.

The transcendental equation (4.20) can be written in terms of the dielectric penetrability tensor. It is enough to use equation (4.3) and make the transformation (4.6), (4.7) accounting for the definition (4.8). One gets [4.3]

$$\varepsilon^l(\omega, k) = 1 + \sum_a \frac{\omega_{L_a}^2}{k^2 v_{T_a}^2}\left[1 - J_+\left(\frac{\omega}{kv_{T_a}}\right)\right] = 0\,, \tag{4.21}$$

where

$$J_+(x) = x\,e^{-x^2/2}\int_{i\infty}^{x} d\tau\,e^{\tau^2/2}\,. \tag{4.22}$$

Low frequency oscillations, where the ions' influence is dominant, are localized in the region $kv_{T_i} \ll \omega \ll kv_{T_e}$. In the case of weak damping the dispersion equation (4.2) can be simplified by means of the asymptotic representation for J_+ in the interval $|x| \ll 1$ for the electron contribution

$$J_+(x) = -ixe^{-x^2/2}\sqrt{2\pi},$$

and by

$$J_+(x) = -ix\sqrt{\pi/2}$$

in the region $|x| \gg 1$, $|\text{Re}\{x\}| \gg |\text{Im}\{x\}|$ for the ion part of the sum in (4.21) or (4.20). The approximate solution of the dispersion equation gives for the real part of ω'

$$\omega'^2 = \omega_{L_i}^2(1 + 3k^2 r_{D_i}^2 + 3r_{D_i}^2/r_{D_e}^2)/(1 + k^{-2}r_{D_e}^{-2}). \tag{4.23}$$

For the imaginary part ω'' under the condition $|e| = |e_i|$ we have

$$\omega'' = -\sqrt{\pi/8}\frac{m_i}{m_e}\frac{(\omega')^4}{k^3 v_{T_e}^3}\left[1 + \sqrt{\frac{m_i}{m_e}}\left(\frac{T_e}{T_i}\right)^{3/2}\exp(-\omega'^2/2k^2 v_{T_i}^2)\right]. \tag{4.24}$$

The dependencies $\omega'(k)$ of the Langmuir and ion waves are shown in Fig. 4.1 [4.3]. The region of strong damping are shown by the broken lines. Ion waves are possible only in nonisothermal plasma in which $T_e \gg T_i$ [4.3,4] in the long wavelength range $kr_{D_e} \ll 1$ and the ion wave spectrum (4.23) is a linear function

$$\omega = \pm k\sqrt{\kappa_B T_e(1 + 3T_i/T_e)/m_i}.$$

These oscillations are known as the ion-acoustic waves and are found in collisionless magnetic hydrodynamics like the usual acoustic waves in a neutral gas. It follows from (4.17) and (4.23) that for both types of waves there are waves propagating in opposite directions.

4.1.2 Weakly Inhomogeneous Bounded Plasma

Let us now take into account a plasma that is inhomogeneous (stratified) along the propagation direction k. This means that the main parameters T_a, ω_{L_a}, r_{D_a} are functions of x. Spatial nonuniformity, as previous investigations of hydrodynamic waves have shown, may cause convective energy transfer between layers of different density [4.9]. We assume that the inhomogeneity scale exceeds the length of the longest plasma waves. As in [4.10], we introduce the local parameters of the wave packet and derive the equations for their evolution to a first approximation of small inhomogeneity parameter. As follows from [4.10] the theory may be generalized to the case of propagation directly at an angle to the density gradient if a ray description is used. We continue to note effects of nonlinearity and waveguide propagation but restrict ourselves to the simplest single-mode projection.

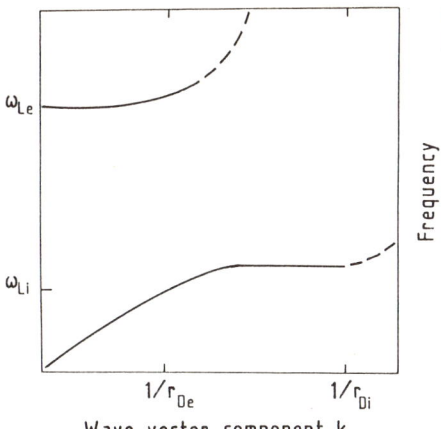

Fig. 4.1. Dispersion curves for ion-acoustic and Langmuir waves

The weak inhomogeneity of plasma ground state with the distribution function $f^0(\beta x)$, $\beta \ll 1$ leads to a corresponding change in the dispersion relation. Let us make the approximate integration up to first order in the inhomogeneity parameter β. By differentiation, the formula for an arbitrary slowly varying function $B(\beta x, \beta t)$ can be tested.

$$\int_0^t B(x - p(t-\tau)/m, \tau) e^{-in\tau} d\tau$$
$$= \left\{ \frac{iB}{n} + n^{-2} \left[B_t + \frac{pB_x(x,t)}{m} \right] \right\} e^{-int} . \qquad (4.25)$$

We neglect higher terms in β and set $\beta = 1$. It is supposed as before that A in (4.11) as well as the derivatives are equal to zero at $t = 0$. Equation (4.14) in this approximation contains the first derivatives with respect to the propagation direction and t:

$$ikA + A_x = 4\pi e^2 \left\{ A \int dp \frac{f_p^0}{i(\omega - pk/m)} - \frac{A}{m} \int \frac{dpp f_{px}^0}{(kp/m - \omega)^2} \right.$$
$$\left. - \int \left[\frac{f_p^0 A_t}{(kp/m - \omega)^2} + \frac{f_p^0 p A_x}{m(kp/m - \omega)^2} \right] dp \right\} . \qquad (4.26)$$

The equality (4.26) represents the approximate evolution equation for the envelope amplitude A. The term that contains f_{px}^0 may be related to the local dispersion relation as well as to the equation for A. In any case it determines the relaxation process which may be associated with convective transport. In the approximation of the Boltzmann distribution and with the condition of $\omega/k \gg v_T$ taking into account (4.16) we get

$$\int \frac{d\mathbf{p} f_p^0}{(kp/m - \omega)^2} = -\frac{2kn}{m\omega^3},$$

$$\frac{1}{m}\int \frac{f_{px}^0 p\, dp}{(kp/m - \omega)^2} = -\frac{N_x}{N}\frac{n}{m\omega^3}, \quad \frac{1}{m}\int \frac{p f_p^0 dp}{(kp/m - \omega)^2} = -\frac{n}{\omega^2 m},$$

where $N = n(x)/[2\pi m \kappa_B T(x)]^{3/2}$. Equation (4.26) after the substitution of the formulas for momentum integrals has the form

$$\frac{2\omega_L^2 k}{\omega^3} A_t + A_x\left[1 - \frac{\omega_L^2}{\omega^2}\right] - \frac{\omega_L^2}{\omega^2}\frac{N_x}{N} A = 0. \tag{4.27}$$

Such a description takes the collisionless Cherenkov-Landau relaxation into account in the dispersion relation for complex ω (4.19,24) and the influence of the inhomogeneity is in the last term in (4.27). If, for example, $N_x/N = \gamma$ does not depend on x and $\omega \approx \omega_1$, then (4.27) is simplified to

$$A_t + c_g A_x - \gamma A = 0, \tag{4.28}$$
$$A = e^{\gamma t}\varphi(x - c_g t), \tag{4.29}$$

where c_g is the group velocity of the wave. If $\gamma > 0$ the density of the plasma increases or the temperature decreases along the wave propagation direction and the wave amplitude in the solution to the Cauchy problem (4.29) grows along the characteristic. When $\gamma < 0$ the opposite effect is observed.

As a boundary condition for a wave disturbance of a plasma state the condition of charged particles can be used [4.3]. The flux of particles or the degree of recombination can be assigned at the boundaries [4.3,11]. The presence of boundaries determines the possible discrete values of k and ω in their spectra as well as the appearance of boundary waves. If the wavelength is much smaller than the interval between the boundaries the discrete evolution equations have an approximate continuous analogue [4.12,13].

Let us note in addition that if the plasma density decreases in some direction upon leaving the hydrodynamical region or, on the other hand, if the contribution of the collision integral increases upon transition to the collisionless regime, then a transitional Knudsen regime interval appears. This requires special technique. Several approaches are discussed in Sect. 5.4.

4.2 Interaction of Plasma Waves

4.2.1 Langmuir and Ion Wave Interactions

Under the action of coherent sources in a plasma the wave trains of different types can be excited. These are the ion and Langmuir longitudinal waves discussed in Sect. 4.1 and waves with $\mathcal{E}$ orthogonal to $\mathbf{k}$. Wave splitting in a simple way is impossible if the plasma is in an external magnetic field, that is, magnetized [4.3]. In the plasma new important dispersion relation branches appear [4.1,3,14].

All possible plasma waves of sufficient intensity interact [4.15,16]. Let us consider the occurrency of an interaction in the case of longitudinal waves in an inhomogeneous plasma. To the end we derive system of equations that describes the Langmuir and ion-acoustic mode interaction. The weak nonlinear dependence of the susceptibility on the electromagnetic field can be found by iterating (4.4) over the amplitude, as was done for the density matrix in Sect. 3.3. To do this, one must change nonlinear Vlasov term $f \to f_0 + \chi$. As χ is a linear function of the amplitude the expression for the distribution function will be quadratic in the field $\mathcal{E}' = \sigma \mathcal{E}$. So if

$$f = f^0 + \sigma \chi + \sigma^2 \Phi + \ldots , \tag{4.30}$$

then

$$\frac{d\Phi}{dt} = -e\mathcal{E}_x \int_0^t F_p\left(x - \frac{p(t-\tau')}{m}, \tau'\right) d\tau' ,$$

where F is determined via (4.9). Since effects of the dispersion and inhomogeneity of the field are assumed to be small ($\beta \ll 1$) during the calculation of Φ we retain only contributions $\sim \beta^0$. Then

$$\Phi = -e^2 \int_0^t \frac{Ae^{i\varphi} + \text{c.c.}}{\omega - pk/m} \left(f_{pp}^0 - \frac{kf_p^0}{m\omega - pk}\right)(Ae^{i\varphi}/i + \text{c.c.})d\tau .$$

Here $\varphi = kx - pkt/m + (kp/m - \omega)\tau$. Integration over t for a uniform plasma gives

$$\Phi = e^2 \frac{(A^2 e^{2i\varphi} + \text{c.c.})[f_{pp}^0 - kf_p^0/(m\omega - pk)]}{2(\omega - pk/m)^2} . \tag{4.31}$$

This contribution to the nonlinearity describes the generation at multiple frequencies and may appear in the propagation of ion-acoustic waves in the range of linearity of the dispersion law (Fig. 4.1) [4.17,18]. The equation for the monodirectional wave may be obtained if in the equation

$$\text{div } \mathcal{E} = 4\pi \sum_a e_a \int d\mathbf{p}[\chi^a + \Phi^a] \tag{4.32}$$

we substitute (4.31) and take into account only the resonant terms adding to relation (4.27) the nonlinear terms proportional to Φ^i. We get

$$A_t^2 + c_{gi}A_x^2 + \frac{1}{2}\frac{d^2\omega}{dk^2}A_{xx}^2 + N_{2i}(A^1)^2 = 0 , \tag{4.33}$$

where $A^{1,2}$ are the amplitudes of the first and second harmonics and the nonlinear constant is proportional to

$$\frac{2\pi e_i^3}{\omega^2} \int \frac{d\mathbf{p}[f_{pp}^{0i} - kf_p^{0i}/(m_i\omega - pk)]}{(1 - pk/\omega m_i)^2} .$$

A rather simple expression for the nonlinear constant can be written if the decomposition (4.15) in the integrand is used. The latter is possible if $\omega/k \gg v_{T_i}$, which occur in an ion-acoustic region which is strongly nonisothermal $T_e \gg T_i$. At an arbitrary ratio of the ion wave phase velocity to the thermal velocity the integral over p can be taken by the transformation to the standard integral $J_+(x)$, defined by (4.22).

In the next approximation, interaction of Langmuir and ion-acoustic waves can appear. For the derivation it is sufficient to take into account the next order in β when χ is calculated. The contribution gives

$$e^2 \mathcal{E}_x \left[\frac{f_p^0 A_t + p(A f_p^0)_x/m}{(k_p/m - \omega)^2} + \text{c.c.} \right]_p . \tag{4.34}$$

Keeping the terms that are in resonance with long ion waves, let us write the equation for the right-hand ion-acoustic wave with slowly varying amplitude Π

$$\Pi_t + c_{\text{gi}} \Pi_x = -N_{\text{ie}} |A|^2 . \tag{4.35}$$

This equation can be obtained if for the electric field components

$$\text{rot rot } \mathcal{E} + \mathcal{E}_{tt}/c^2 = -4\pi \mathbf{j}_t/c^2$$

one includes the current density as the functional of the distribution function (4.3)

$$\mathbf{j} = \sum_a e_a \int d\mathbf{p}_a \mathbf{p}_a (\sigma \chi^a + \sigma^2 \Phi^a)/m_a ,$$

and projects the result on the longitudinal direction (x-axis). Φ can be calculated to the next order in σ if only the mean field with frequency $\omega \sim 0$ is retained in (4.34). We write the corresponding contribution in the electron function (4.30) denoting it as

$$\Phi_{\text{int}}^e = -\frac{e^2}{\omega^2} \int \Biggl\{ |A(x - p(t-\tau)/m, \tau_1)|_{\tau_1}^2 \Bigg|_{\tau_1 = \tau}$$

$$\times \left[\frac{f_p^0}{(1 - kp/m\omega)^2} \right]_p + |A|_x^2 \left[\frac{f_p^0 p/m}{(1 - kp/m\omega)^2} \right]_p$$

$$+ |A|^2 \left[\frac{f_{xp}^0 p/m}{1 - pk/m\omega} \right]_p \Biggr\} d\tau . \tag{4.36}$$

The equations of the first approximation (4.28) can be read as $A_t \simeq c_g A_x$ and the dependence of $A(\beta(x - c_g t), \beta^2 t)$ on the arguments is weak. To the first order in β the derivatives of A with respect to τ and τ_1 differ only by factor $A_\tau = (p/m - c_g) A_x$, $A_{\tau_1} = A_\tau (1 - p/m c_g)$. Integrating over τ and the momenta we get the nonlinear term in (4.32) and therefore also in (4.35):

$$\int d\boldsymbol{p}\Phi^e_{\text{int}} = -\frac{e^2}{\omega^2}|A|^2 \int \frac{f_p^0/mc_g d\bar{p}}{(1-p/mc_g)(1-kp/m\omega)^2}\,.$$

In the region $\omega/k \gg v_{T_e}$, $c_g \gg v_{T_e}$,

$$\begin{aligned}4\pi e \int d\boldsymbol{p}\Phi^e_{\text{int}} &= \frac{4\pi e^2 n}{m\omega^2}\frac{1}{mc_g}\left(\frac{1}{c_g}+\frac{2k}{\omega}\right)|A|^2 \\ &= \frac{\omega_{L_e}^2}{\omega^2}\frac{\omega+2kc_g}{c_g}|A|^2\,.\end{aligned}$$

One should remember that here ω is the carrier frequency of the Langmuir wave train $\omega \approx \omega_{L_a}$.

In the same manner the nonlinear term in the equation for the amplitude function of the Langmuir wave can be derived. The term is obviously proportional to A multiplied by Π_x. The equation for the ion-acoustic wave is often written in the notations of the plasma density n [4.19]. The density of plasma is proportional to the derivative of Π with respect to x according to (4.2). Therefore in the equation for the density perturbation the term $|A|^2_x$ will stand on the right-hand side. The system of equations for interacting plasma waves can be transformed to the equation containing both left and right ion waves.

$$n_{tt} - n_{xx} = |A|^2_{xx}\,,$$

and to the equation for the Langmuir wave amplitude

$$2\mathrm{i}A_t + 3A_{xx} - nA = 0\,. \tag{4.37}$$

The scale has been transformed to simplify the coefficients. The system (4.37) has been called the Zakharov system [4.19]. The investigation of (4.37) shows the Lyapunov stability of the stationary solutions and the existence of the mass, energy and plasmon number conservation laws [4.20]. The two-dimensional generalization of (4.37) has unstable (by Lyapunov) stationary solutions [4.20]. The interaction of the Langmuir stationary waves with plasma particles was considered in [4.21].

4.2.2 Three-Wave Helico-Acoustic Interaction

The other limiting description of an electromagnetic field interacting with plasma is given by the hydrodynamical approximation. In this description with Knudsen number $\mathrm{Kn} \ll 1$, the collision integral plays the fundamental role and the plasma is considered as a conductive fluid. The distribution is described by a local equilibrium function. The condition for such an approach to be valid is the smallness of the mean free path of plasma particles compared to all the scales of the wave disturbance and the wave frequencies compared with the collision frequencies [4.3,5,9]. The method of the nonlinear wave theory in this case is the same as described in Chaps. 2,3 but is more complex because the number of dynamical variables increases with the number of medium components.

If the region of localization of the plasma particles that is related to the thermal motion or the Larmour rotation is smaller than the minimum wave dimensions the equations of free particle motion are used. The integration of the equations of motion of the charged particle in an electromagnetic and any additional (e.g., gravitational) field allows one to find the particles velocity that determines a displacement current density j. The obtained mass equations are completed by the hydrodynamical system. This allows one to derive the dispersion relation and to generalize the theory to the nonlinear case [4.6].

The assumptions in the theory of electrodynamics are quite similar to those made in multi-stream hydrodynamics [4.9]. This approximate description may be based on the self-consistent Vlasov equation if the distribution function is taken in the form

$$f_a(p, r, t) = n_a(r, t)\delta(p - m_a u_a(r, t)) \ .$$

We neglect here the thermal momentum fluctuations as in the mechanical approach. The system of equations is

$$\frac{\partial n_a}{\partial t} + \mathrm{div}\,(n_a u_a) = 0 \ ,$$
$$\frac{\partial u_a}{\partial t} + (u_a \nabla) u_a = \frac{e_a}{m_a}\left\{\mathcal{E} + \frac{1}{c}[u_a \times B]\right\} + g \ . \tag{4.38}$$

Here g denotes the mass density of the nonelectromagnetic force. The field equations are simplified as well:

$$\mathrm{div}\,\mathcal{E} = 4\pi \sum_a e_a n_a, \quad \mathrm{rot}\,B = \frac{1}{c}\frac{\partial \mathcal{E}}{\partial t} + \frac{4\pi}{c}\sum_a e_a n_a u_a \ . \tag{4.39}$$

Equations (4.38) are obviously nonlinear. Subsequent approximations in the amplitude and dispersion parameters in the operators of these equations give the necessary equalities that lead to the equations of the weakly nonlinear theory.

We shall show this as an example. A cold plasma is placed into a constant magnetic field. Any motion in the plasma then depends strongly on this field. A tendency towards a circular motion with the characteristic Larmour frequency $\Omega_a = e_a B_0/m_a c$ appears [4.6]. Let the intensity of the field be not large: $\Omega_e^2 \gg \omega \gg \sqrt{|\Omega_e \Omega_i|}$. Then the influence of the ions in the transverse waves may be neglected [4.4]. We suggest also that the time be measured in units Ω_e^{-1}. The right-hand side in the equation of motion (4.38) then will be much larger than the left-hand side because the time derivatives are small due to the relatively small frequency and because the convective term is quadratic in amplitude. Moreover, in this approximation we shall linearize the right side by the magnetic field

$$\mathcal{E} + \frac{1}{c}[u_e \times B_0] = 0 \ . \tag{4.40}$$

Vectorially multiplying (4.40) by B_0 and taking into account that $(B_0, u_e) \approx 0$, we find

$$\boldsymbol{u}_e = -e[\boldsymbol{B}_0 \times \mathcal{E}]/B_0^2 \,. \tag{4.41}$$

Equation (4.39) after the substitution of (4.41)

$$\text{rot}\,\boldsymbol{B} = -4\pi e n_e [\boldsymbol{B}_0 \times \mathcal{E}]/B_0^2 \,, \tag{4.42}$$

together with $-\boldsymbol{B}_t/c = \text{rot}\,\mathcal{E}$, form the closed system that can be transformed to the form of (4.36)

$$-\Delta \mathcal{E} = 4\pi e n_e [\boldsymbol{B}_0 \times \mathcal{E}_t]/cB_0^2 \,, \tag{4.43}$$

since $\text{div}\,\mathcal{E} = 0$. Taking the Fourier transformation we get the dispersion relation $\omega = (cB_0/4\pi|e|n_e)k^2$ and the polarization coupling $\mathcal{E}_x = -i\mathcal{E}_y$. Such waves in a dense plasma are called helicoidal (or helicon quasiparticles). In ionospheric plasma physics such oscillations are interpreted as being the "whistling atmospherics" (spiral waves) [4.3].

The fundamental mode of the magnetoactive plasma is the transverse wave with $\omega \ll \Omega_i$. Its spectrum is given by the function $\omega^2 = k^2 v_A^2$ where $v_A = B_0/\sqrt{4\pi n_i m_i}$. These waves are the Alfven or MHD (magneto-hydrodynamic) waves. In a magnetoactive plasma, longitudinal or ion-acoustic waves are also present.

Let us consider the electromagnetic modes interaction in a magnetoactive plasma. The interaction is determined by terms that appear in (4.42,43) if (4.41) is taken to the next higher order. To do this we put (4.40) into the convective term of (4.38)

$$(\boldsymbol{u}_e, \nabla)\boldsymbol{u}_e = -c^2 \left([\boldsymbol{B}_0 \times \mathcal{E}], \nabla\right)[\boldsymbol{B}_0 \times \mathcal{E}]/B_0^4$$
$$= -c^2 [\boldsymbol{B}_0 \times ([\boldsymbol{B}_0 \times \mathcal{E}], \nabla)\mathcal{E}]/B_0^4 \,.$$

Solving again (4.38) for u_a in the right-hand side we write the quadratic contribution to the electron velocity

$$\boldsymbol{u}_e^{(2)} = \frac{mc^3}{eB_0^6}([\boldsymbol{B}_0 \times \mathcal{E}], \nabla)\,\mathcal{E} - \frac{mc^3 \boldsymbol{B}_0}{eB_0^6}\left(\boldsymbol{B}_0([\boldsymbol{B}_0, \times \mathcal{E}], \nabla)\mathcal{E}\right) \,. \tag{4.44}$$

Substituting (4.41) and (4.44) into the expression for the current density and the result in (4.3), we get the nonlinear analogue of (4.42) and as a corollary (4.43).

We now represent the electric field as a superposition of three wave packets with amplitudes $\mathcal{E}_1$, $\mathcal{E}_2$, $\mathcal{E}_3$ and group velocities v_1, v_2, v_3 and frequencies ω_i. Let the first train be ion-acoustic and the second and third be helicoidal waves. Then, using the polarization and dispersion relations we go to the standard three-wave system (Sects. 2.7, 3.3, [4.22]):

$$\left(\frac{\partial}{\partial t} + v_i \frac{\partial}{\partial x}\right) \mathcal{E}_i = \beta_i \mathcal{E}_j \mathcal{E}_k \,, \tag{4.45}$$

$i \neq j \neq k$. The coefficients of the nonlinear coupling β_i (3.52) are equal to [4.22]

$$\beta_1 = e\omega_{L_e}^2/4\varepsilon_0 v_\phi \omega_2 \omega_3 m^*$$
$$\beta_2 = \beta_3 = e\omega_{L_e}^2/2m^* v_\phi \omega_1 \omega_3 \left[\mathcal{E}_0 + (\omega_{L_e}/\Omega_e)^2\right] ,$$

where ε_0 is the static dielectric constant, v_ϕ is the phase velocity of the ion-acoustic wave, m^* the effective electron mass.

This type of interaction is applicable to problems in ionosphere and solid state physics. It is assumed that the system is placed into a magnetic field with the vector B_0 parallel to the x axis at a temperature $T \ll m^* v_\phi^2/\kappa_B$. When the helicons have equal group velocities, $v_2 = v_3$, the solution of the system is transformed to a problem for self-induced transparency equations. The solution of the latter by inverse methods gives the condition for the appearance of a soliton in the initial three-wave disturbance. The analytical expression for this contribution coincides with (3.53) and the relation between the parameters with (3.54).

Let us give the results of the estimations for a helico-acoustic interaction in n-InSb under weak plasma wave damping. If $T = 77K$, the electron density is equal to $n_0 = 1, 2 \cdot 10^{14}$ cm^{-3}, electron mobility is $3.5 \cdot 10^5$ cm/V $\cdot$ s, $B_0 = 0.7$T, the frequencies $\omega_{2,3} = 5 \cdot 10^{11}$s^{-1}, $\omega_1 = 3 \cdot 10^7$s^{-1}, phase velocity of a sound wave $v_p = 3 \cdot 10^3$cm/s, relaxation times are $\tau_c = 10^{-11}$s, $m_a^* = 0.014 m_e$, $\varepsilon_0 = 17$. The necessary amplitude of the electric field in a helicoidal wave is about 10^4 V/m and for an acoustic wave, about 10^2 V/m at a pulse duration of $\tau_p = 10^{-6}$s. At such an amplitude a sharp rise in the wave amplitude should occur after passing through the plasma. In contrast, the nonlinear velocity of propagation decreases.

Thus, this method of simplification of the theoretical description of weakly nonlinear plasma disturbances leads to the canonical equations (4.33,35) and (4.45) which are widely discussed in the literature. The introduction to the methodology given in Sect. 4.2,3 for plasma waves should aid in solution of applied problems discussed in Sects. 4.4,5 restricted to the one-dimensional case. Three-dimensional problems which lead to new concepts are discussed in [4.2,14,23].

4.3 Statistical Averaging and Weak Turbulence in Plasma

4.3.1 Spectral Densities of Electromagnetic Field Fluctuations

The repeated iteration procedure in kinetic hydrodynamics approaches leads to a common form of the constitutive equation [4.9]

$$\begin{aligned} D'_i(\bar{r}, t) = \sum_{n=1}^{\infty} &\int_{-\infty}^{t} dt_1 \int d\mathbf{r}, \ldots \int_{-\infty}^{t_{n-1}} dt_n \int d\mathbf{r}_n \\ &\times \varepsilon_{ij(1)\ldots j(n)}(t - t_1, \mathbf{r} - \mathbf{r}_1; \ldots t_{n-1} - t_n, \mathbf{r}_{n-1} - \mathbf{r}_n; t_n, \mathbf{r}_n) \\ &\times \mathcal{E}_{j(1)} \ldots \mathcal{E}_{j(n)}(t_n, \bar{r}_n) \end{aligned} \quad (4.46)$$

which is the nonlinear generalization of (4.6).

For subsequent statistical averaging over the stochastic phase, the equation for a quadratic combination of slowly varying amplitudes $\mathcal{E}_i(\omega, k, \bar{r}, t)$ is derived. Substituting (4.46) into the Poynting equation and averaging over a time interval that is much longer than the wave periods $2\pi\omega^{-1}$ one gets

$$\frac{\partial}{\partial t}\left[\frac{\partial(\omega M_{ij})}{\partial \omega}\mathcal{E}_i^*(\omega, k)\mathcal{E}_j(\omega, k)\right] - \frac{\partial}{\partial r_s}\left[\frac{\partial(\omega M_{ij})}{\partial k_s}\mathcal{E}_i^*(\omega, \bar{k})\mathcal{E}_j(\omega, \bar{k})\right]$$
$$= i\omega \sum_{n=1}^{\infty}\int d\omega_1 \ldots d\bar{k}_{n-1} \varepsilon_{ij(1)\ldots j(n)}(\omega, k, \omega_1, k_1; \ldots; \omega_{n-1}, k_{n-1})$$
$$\times \mathcal{E}_i^*(\omega, k)\mathcal{E}_{j(1)}(\omega - \omega_1 k - k_1)\ldots \mathcal{E}_{j(n-1)}(\omega_{n-2} - \omega_{n-1}, k_{n-2} - k_{n-1})$$
$$\times \mathcal{E}_{j(n)}(\omega_{n-1}, k_{n-1}) + c.c. \tag{4.47}$$

Here $B = c[k \times \mathcal{E}]/\omega$, the symmetry of the nonlinear polarization constants is as in (4.46) and the identity $\mathcal{E}(-\omega, -k) = \mathcal{E}^*(\omega, k)$ holds. Also $\varepsilon_{ij}^H = 1/2(\varepsilon_{ij} + \varepsilon_{ji}^*)$ and

$$M_{ij} = \varepsilon_{ij}^H(\omega, k) + \frac{c^2 k^2}{\omega^2}\left(\delta_{ij} - \frac{k_i k_j}{k^2}\right).$$

Averaging over the statistical ensemble of waves that differ by an accidental phase is denoted by the symbol $\langle \rangle$. It is assumed that a stationary point and spatial homogeneity leads to

$$\langle \mathcal{E}(\omega, k)\rangle = 0$$
$$\langle \mathcal{E}_i(\omega, k)\mathcal{E}_j^*(\omega', k')\rangle = \delta(\omega - \omega')\delta(k - k')(\mathcal{E}_i\mathcal{E}_j)_{\omega, k}, \tag{4.48}$$

where $(\mathcal{E}_i\mathcal{E}_j)_{\omega, k}$ is the spectral density of the electric field fluctuations. It follows from (4.48) and the linear equations for spectral functions obtained from (4.47), which expresses the approximate absence of correlations between functions with different ω, k that the correlators of third order are equal to zero, and the fourth order correlators are expressed through the spectral functions. In the approximation that the field amplitude is quadratic the spectral transform of the field equations gives

$$\mathcal{E}_i(\omega, k) = \mathcal{E}_i^0(\omega, \bar{k}) - M_{ij}^{-1}\int d\omega\, dk_1 \varepsilon_{ij(1)j(2)}(\omega, k, \omega_1, k_1)$$
$$\times \mathcal{E}_{j(1)}(\omega - \omega_1, k - k_1)\mathcal{E}_{j(2)}(\omega_1, k_1). \tag{4.49}$$

If one substitutes (4.49) into the third and fourth order correlators and also neglects the self-action then the averaging of (4.47) yields the generalized kinetic field equation [4.9]

$$\frac{\partial}{\partial t}\left\{\left[\frac{\partial[\omega\varepsilon_{ij}^H(\omega,\boldsymbol{k})]}{\partial\omega}+\frac{c^2k^2}{\omega^2}\left(\delta_{ij}-\frac{k_ik_j}{k^2}\right)\right]\frac{1}{\omega}(\mathcal{E}_j\mathcal{E}_i)_{\omega,\boldsymbol{k}}\right\}$$

$$+\frac{\partial}{\partial r_s}\left\{\left[-\frac{\partial[\omega\varepsilon_{ij}^H(\omega,\boldsymbol{k})]}{\partial k_s}+\frac{c^2}{\omega^2}(2k_s\delta_{ij}-k_j\delta_{si}-k_i\delta_{sj})\right]\frac{1}{\omega}(\mathcal{E}_j\mathcal{E}_i)_{\omega,\boldsymbol{k}}\right\}$$

$$=2i\mathcal{E}_{ij}^a(\omega,\boldsymbol{k})(\mathcal{E}_j\mathcal{E}_i)_{\omega,\boldsymbol{k}}+\mathrm{Im}\int d\omega'\,d\boldsymbol{k}'\left[M_{ij}^{*-1}(\omega,\boldsymbol{k})S_{ij(1)j(2)}(\omega,\boldsymbol{k},\omega',\boldsymbol{k}')\right.$$

$$\times S_{jj(3)j(4)}(\mathcal{E}_{j(2)}\mathcal{E}_{j(4)})_{\omega',\boldsymbol{k}'}(\mathcal{E}_{j(1)}\mathcal{E}_{j(3)})_{\omega-\omega';\boldsymbol{k}-\boldsymbol{k}'}+2M_{j(1)j}^{-1}(\omega-\omega',\boldsymbol{k}-\boldsymbol{k}')$$

$$\times S_{ij(1)j(2)}S_{jj(4)j(3)}(\omega-\omega',\boldsymbol{k}-\boldsymbol{k}',\omega,\boldsymbol{k})(\mathcal{E}_{j(2)}\mathcal{E}_{j(4)})_{\omega'\boldsymbol{k}'}(\mathcal{E}_{j(3)}\mathcal{E}_i)_{\omega,\boldsymbol{k}}$$

$$\left.-2(\mathcal{E}_{j(1)}\mathcal{E}_i)_{\omega,\boldsymbol{k}}(\mathcal{E}_{j(3)}\mathcal{E}_{j(2)})_{\omega',\boldsymbol{k}'}V_{ij(2)j(1)j(3)}(\omega,\boldsymbol{k},\omega',\boldsymbol{k}')\right],\qquad(4.50)$$

where

$$V_{ij(2)j(1)j(3)}=\varepsilon_{ij(2)j(1)j(3)}(\omega,\boldsymbol{k};\omega+\omega',\boldsymbol{k}+\boldsymbol{k}';\omega',\boldsymbol{k}')$$
$$+\varepsilon_{ij(2)j(3)j(1)}(\omega,\boldsymbol{k};\omega+\omega',\boldsymbol{k}+\boldsymbol{k}';\omega,\boldsymbol{k}),$$
$$S_{ij(1)j(2)}=2\varepsilon_{ij(1)j(2)}(\omega,\boldsymbol{k};\omega',\boldsymbol{k}').$$

The equations for the averages (4.50) (the turbulent spectral densities) allow one to describe quantitatively turbulent processes in weakly nonuniform plasma which is related to the parametric waves interaction. The theory of parametrical turbulence in a spatially uniform plasma is discussed in [4.24] and the generalization to the spatially nonuniform case is suggested in [4.9]. An alternative approach is reported in [4.25].

Nonlinear interactions of waves in all probability play an important role in the evolution of turbulent states. In the current theory of weak turbulence the generating mechanism is a nonlinear instability of "explosive" type. The appearance of ordered structures is associated with stable or metastable states of the nonlinear systems (solitons or quasisolitons) [4.23]. As follows from the analysis of (4.33,35), in plasma an aperiodic process may develop that leads to an intense exchange of energy between degrees of freedom. The Langmuir collapse is a characteristic example of such a process [4.19,23], and is described by the three-dimensional equation system that is the generalization of (4.37). Investigation of solutions of the Zakharov equations shows that a three-dimensional initial disturbance of partial stationary solutions leads to the development of the instability.

In [4.9] it is shown that a convective energy transfer in the direction of the plasma density gradient effectively suppresses the aperiodic parametric instability. The approach based on the system (4.50), generalized for a weakly inhomogeneous plasma is developed. The main necessary assumptions for the generalization are described in Sect. 4.1. It is assumed that there exists a region, with dimensions on the order of the dimensions of the inhomogeneity β^{-1}, in which the conditions of the parametric resonance are fulfilled. In this region the wave can grow exponentially. Thus, let us allow the instability to be saturated by the decay interaction of Langmuir and ion-acoustic waves. The full kinetic equations of the statistical theory of plasma turbulence for Langmuir numbers

$N_L^\sigma(\omega, \boldsymbol{k}_1, x)$ and ion-acoustic $N_S^\sigma(\omega, \boldsymbol{k}_\perp, x)$ waves are normalized so that the averaged electric field for every mode is given by the expression

$$\frac{\mathcal{E}_{L,S}^2}{8\pi} = \frac{\int d\omega\, \omega dk_1 [N_{L,S}^+(\omega, \boldsymbol{k}_1, x) + N_{L,S}^-(\omega, \boldsymbol{k}_1, x)]}{(2\pi)^3 v_g(\omega, \boldsymbol{k}, x)}, \quad v_g = \left[\frac{\partial k_x}{\partial \omega}\right]^{-1}.$$

The integration is done over the spectrum of the corresponding dispersion branch. For $\mathcal{E}_L$ this is $\omega > \omega_{L_e}(x)$, and for $\mathcal{E}_S$ over $0 < \omega < \omega_{L_i}$ (Fig. 4.1).

4.3.2 One-Dimensional Turbulence

Let us consider as in [4.9] the one-dimensional turbulence, assuming that the main cause of the instability is an inhomogeneity along the x direction. Substituting $N^\sigma(\omega, \boldsymbol{k}_\perp, x) = (2\pi)^2 \delta(\boldsymbol{k}_\perp) N^\sigma(\omega, x)$ into the kinetic equations we get for a stationary state

$$\sigma v_L(\omega) \frac{\partial}{\partial x} N_L^\sigma(\omega) = -2\gamma_L^\sigma N_L^\sigma(\omega)$$
$$+ \Gamma[P^{-\sigma}(\omega + k_L v_S) - P^\sigma(\omega - k_L v_S)],$$
$$\sigma v_S \frac{\partial}{\partial x} N_S^\sigma(\omega) = -2\gamma_S^\sigma(\omega) N_S^\sigma + \Gamma P^\sigma(\omega_L(\omega, x), x), \quad (4.51)$$

where the function $P^\sigma(\omega, x)$ is determined by

$$P^\sigma(\omega) = N_L^\sigma(\omega + k_L(\omega) v_S) N_L^{-\sigma}(\omega - k_L(\omega) v_S)$$
$$+ N_S^\sigma(2k_L(\omega) v_S) [N_L^\sigma(\omega + k_L(\omega) v_S) - N_L^{-\sigma}(\omega - k_L(\omega) v_S)].$$

Here $\Gamma = \omega_{L_e}^2 \omega_{L_i}/24 n_e T_e v_{T_e}$ is the nonlinear interaction constant. The wavenumber $k_L(\omega)$ and ω_L at $\boldsymbol{k}_\perp = 0$ are determined from the dispersion equation (4.17)

$$\omega_L(\omega, x) = \left[\omega_{L_e}^2(x) + \frac{3 v_{T_e}^2 \omega^2}{4 v_S^2}\right]^{1/2}$$

$$k_L(\omega, x) = \left[\frac{\omega^2 - \omega_{L_e}^2(x)}{3 v_{T_e}^2}\right]^{1/2}$$

and for the sound waves, as a result of (4.23) $k_x = k_S = (\omega/v_S)^2$, $v_S = v_{T_e} \omega_{L_i}/\omega_{L_e}$ where $\gamma_{S,L}^\sigma$ are the increments of increase (decrease) for corresponding waves. If instead of x we introduce the new variable $\Omega = 2k_L(\omega, x) v_S$ which is equal to the frequency of the sound wave which is excited during the decay of the Langmuir wave with a frequency ω at x, then $L^{-1} = -\ln x\omega_{L_e}^2 = \text{const}$.

Through a scale transformation we can eliminate the nonlinear constants:

$$l^\sigma(\omega, \Omega) = \frac{2\Gamma L}{v_S} N_L^\sigma(\omega, x)$$

$$s^\sigma(\omega, \Omega) = \frac{2\Gamma L}{v_S} N_S^\sigma(2k_L(\omega, x) v_S, x).$$

As in Sects. 4.1,2, because the plasma is assumed to be cold and nonisothermic, $\omega_{L_i} \ll k_L v_{T_e} \ll \omega_{L_e}$, we go to the system

$$2\sigma\omega \frac{\partial l^\sigma}{\partial \Omega} = -2\lambda_L^\sigma l^\sigma + l^{-\sigma}(\omega+\Omega)l^\sigma(\omega) + s^\sigma\left(\omega+\frac{\Omega}{2}\right)$$
$$\times \left[l^\sigma(\omega+\Omega) - l^{-\sigma}(\omega)\right] - l^\sigma(\omega)l^{-\sigma}(\omega-\Omega) - s^\sigma\left(\omega-\frac{\Omega}{2}\right)$$
$$\times \left[l^\sigma(\omega) - l^{-\sigma}(\omega-\Omega)\right], \tag{4.52}$$

$$-\sigma\omega \frac{\partial s^\sigma}{\partial \Omega} = -2\lambda_S^\sigma s^\sigma + l^\sigma\left(\omega+\frac{\Omega}{2}\right)l^{-\sigma}\left(\omega-\frac{\Omega}{2}\right)$$
$$+ s^\sigma(\omega)\left[l^\sigma\left(\omega+\frac{\Omega}{2}\right) - l^{-\sigma}\left(\omega-\frac{\Omega}{2}\right)\right], \tag{4.53}$$

where $\lambda_{L,S}^\sigma = 2L\gamma_{L,S}^\sigma/v_S$ is the dimensionless analogue of $\gamma_{L,S}^\sigma$. Equations (4.52,53) are a system of four nonlinear equations, and the pair of equations (4.53), for the spectral densities of acoustic waves, is linear if the amplitudes l^σ are known. Therefore for the solution of the system it is natural to use a splitting method, first integrating the expression for the Langmuir waves and then finding the sound wave and iterating further.

The boundary condition for the system is chosen so that at the turning line $\Omega = 0$ the numbers of quanta of the approaching and leaving Langmuir waves are equal

$$l^+(\omega, 0) = l^-(\omega, 0). \tag{4.54}$$

This condition holds when the turbulence level in a neighborhood of $\Omega = 0$ is small, which is possible if $\Gamma N_L \ll \omega_{L_i}(\delta\omega/\omega_{L_e})^{1/2}$ or $l(\omega, 0) \lesssim (\delta\omega/\omega_{L_e})^{1/2} L/r_{D_e}$ where $\delta\omega$ is a frequency width of the turbulence spectrum.

The second boundary condition for the Langmuir mode is determined by the Landau damping at large Ω:

$$l^-(\omega, \Omega)\big|_{\Omega \gg \Omega_m} \longrightarrow 0. \tag{4.55}$$

The turbulence region is restricted along the ω axis ($\omega \sim \omega_0$ is the frequency of the excitation field). Therefore it is required that the amplitude of a sound wave that is propagated into the region of turbulence decreases

$$s^+(\omega \gg \omega_0, \Omega) \longrightarrow 0, \quad s^-(\omega \ll \omega_0, \Omega) \longrightarrow 0. \tag{4.56}$$

The process of increasing plasma turbulence may be described as follows. The external plasma field in the region of parametric amplification transfers energy to one of the Langmuir modes that is distributed between other modes due to the interaction. The formation of a quasistationary turbulence spectrum includes the processes of transfer in a decay interaction as well as the convective energy exchange [4.9].

In the case of strong ion-acoustic wave damping we can separate the system (4.52,53). To do this we neglect the influence of the sound wave on the energy flux along the spectrum and throw out terms with s^σ in (4.52). Then

$$2\sigma\omega \frac{\partial l^\sigma}{\partial \Omega} = l^\sigma(\omega)\left[l^{-\sigma}(\omega + \Omega) - l^{-\sigma}(\omega - \Omega)\right]. \tag{4.57}$$

Following a method similar to the one discussed in Sects. 2.5,6, we note that in the zeroth approximation the quadratic of the Langmuir spectral density and the damping terms are the most important in (4.53) for s^σ. Then s^σ can be found in the obvious form

$$s^\sigma = (2\lambda_S)^{-1} l^\sigma \left(\omega + \frac{\Omega}{2}, \Omega\right) l^{-\sigma}\left(\omega - \frac{\Omega}{2}, \Omega\right).$$

The authors of [4.9] relied on the condition that the width of the spectrum turbulence exceeds the transfer step ($\delta\omega > \Omega$). They adopt a continuum approximation for the displacement operator in the right-hand side of (4.57) and change $l(\omega + \Omega) - l(\omega - \Omega) \approx 2\Omega \partial l(\omega)/\partial\omega$. Introducing $\psi(y,z) = l^+(\omega,\Omega)l^-(\omega,\Omega)$, $y = \omega^2$, $z = \Omega^2$, one gets

$$\psi_{yy} + \ln_{zz}\psi = 0. \tag{4.58}$$

The particular solution that satisfies the boundary conditions (4.54,55) may be obtained by separation of variables

$$l^\sigma = \left[\frac{\omega^2 - \omega_m^2}{\Omega_m^2}\right]\left[1 + \sigma \tanh\frac{\Omega^2}{\Omega_m^2}\right]. \tag{4.59}$$

The integration constant here is the lowest boundary of the turbulence spectrum, Ω_m is the dimension of the spectrum change along Ω. The derived equation and its solutions allow one to relate the turbulence spectrum at the edge of the amplification region ω_α and within the interval $(\omega_\alpha, \omega_m)$. One can also calculate the energy flux that is transported by Langmuir waves from the parametric absorption area to the Cherenkov damping region.

The solution (4.59), as the authors of [4.9] note, is only qualitative because it is obtained for the special case of spectrum at $\omega = \omega_\alpha$. However, the behavior of the solution is typical at realistic increments. The weak dependence of a characteristic turbulence level on Ω allows one to estimate the order of the width of a Langmuir turbulence spectrum $\delta\omega = \omega_\alpha - \omega_m$. Furthermore, the obtained equations and approximations allow the study of Cherenkov electron heating within a framework of a quasilinear theory. The deformation of the electron distribution function in the diffusion approximation has also been investigated [4.9].

The described method provides a basis for understanding and estimating the influence of powerful radiation on plasma. One can follow the path of energy and spectral redistribution in the limits of the dynamical theory for longitudinal one-dimensional plasma waves in inhomogeneous media. The listed sequence of the model equations for the spectral properties of plasma waves provides the possibility for developing plasma turbulence theory [4.26]. In the next section another approach to the solution of equations (4.56) is discussed.

4.4 Multisoliton Solutions of Discrete Silin–Tikhonchuk Equations

4.4.1 The Darboux–Matveev Transformation

In this section the theory of the equations (4.57) is developed without a transition to the continuum form (4.58). The equations of interaction of a left and right Langmuir plasma wave with strong sound damping is written by using translation operators on the step function

$$T^\pm f(\omega, \Omega) = f(\omega \pm \Omega, \Omega) ,$$

denoting

$$\frac{1}{2}l^+ = p(\omega, \Omega), \quad \frac{1}{2}l^- = s(\omega, \Omega) ,$$
$$\omega p_\Omega = p(T^+ + T^-)s ,$$
$$-\omega s_\Omega = s(T^+ - T^-)p . \qquad (4.60)$$

The identity $(\partial/\partial\Omega)T^\pm f = T^\pm(\partial f/\partial\Omega) \pm T^\pm(\partial f/\partial\omega)$ holds. If the inhomogeneity in ω is much larger than the scale along the Ω axis, i.e., $|p_\omega| \ll |p_\Omega|$, then the translation operators $T^\pm$ commute with $\partial/\partial\Omega$ and the discrete analogue for (4.60) can be introduced $\omega/\Omega \sim n$,

$$\dot{p}_n = p_n(s_{n+1} - s_{n-1}), \quad -\dot{s}_n = s_n(p_{n+1} - p_{n-1}) . \qquad (4.61)$$

Here $l^+(\omega_\pm\Omega, \Omega) = \tilde{l}^+(\Omega(\omega/\Omega \pm 1), \Omega) = 2p_{n\pm1}(\Omega')$, $\Omega/\omega_{L_e} = \Omega'$, and the dot denotes here and below the derivative with respect to Ω'. Physically, the assumption of discretization implies that the variation in the spectral density is smaller than the scale of the plasma inhomogeneity.

Let us go to the matrix form of (4.61). If

$$K_n = \begin{pmatrix} -p_n & 0 \\ 0 & s_n \end{pmatrix}, \quad \sigma_1 = \begin{pmatrix} 0 & 1 \\ 1 & 0 \end{pmatrix}, \quad R_n = \sigma_1 K_n \sigma_1 K_{n+1} ,$$

then

$$\dot{K}_n = K_n \sigma_1 K_{n+1} \sigma_1 - \sigma_1 K_{n-1} \sigma_1 K_n . \qquad (4.62)$$

For this equation there exists an L-A pair

$$K_n \Psi_{n+1} + \Psi_{n-1} = \sigma \Psi_n \Lambda ,$$
$$\dot{\Psi}_n = K_n \Psi_{n+2} , \qquad (4.63)$$

where Ψ_n is a matrix function of Ω' and $\Lambda = \begin{pmatrix} 0 & \lambda \\ \mu & 0 \end{pmatrix}$ is the matrix spectral parameter.

The method of the integration of (4.62) is based on Matveev's works and allows one to find a large set of solutions to nonlinear equations by means of algebraic manipulations. It is especially effective for two-dimensional problems [4.27] and for equations with nonlocal dispersion laws [4.28]. This method does not require the inverse problem formulation and therefore helps to avoid the difficulties associated with it [4.12]. The technique is similar to the methods of Hirota and Backlund. The advantage of the Darboux–Matveev transformations is in their straightforwardness. The generalization of the results to matrix equations [4.13] allows one to formulate the following theorem for the system (4.62).

If the Darboux–Matveev transformation is introduced by

$$\Psi_n[1] = \varphi_{n-2}(1)\varphi_n^{-1}(1)\Psi_n - \Psi_{n-2}, \tag{4.64}$$

$$K_n[1] = \sigma_1 \varphi_{n-2}(1)\varphi_n^{-1}(1)\sigma_1 K_n \varphi_{n-1}^{-1}(1), \tag{4.65}$$

where $\varphi_n(1)$ is a solution of (4.63) for the fixed matrix $\Lambda = \Lambda_1$ then the system (4.63) is transformed into itself by the substitution $\Psi_n \to \Psi_n[1]$, $K_n \to K_n[1]$. The proof can be made by direct substitution. The scheme of the proof is contained in [4.13].

Carrying out the transformation (4.64), in addition to new solutions of the discrete Silin–Tikhonchuk equations, we get the new solution of the linear system (4.63) for which (4.62) is the compatibility condition. Consequently, the transformation can be repeated and iterated functions $\Psi_n[N]$ and $K_n[N]$ may be obtained. We write the functions

$$K_n[N] = \sigma_1 \frac{\Delta_{n-2}[N]}{\Delta_n[N]} \sigma_1 K_n \frac{\Delta_{n+1}[N]}{\Delta_{n-1}[N]}, \tag{4.66}$$

where

$$\Delta_n[N] = \begin{vmatrix} \varphi_n(1) & \cdots & \varphi_n(N) \\ \varphi_{n-2}(1) & \cdots & \varphi_{n-2}(N) \\ \cdots & & \\ \varphi_{n-2N+2}(1) & & \varphi_{n-2N+2}(N) \end{vmatrix}. \tag{4.67}$$

The results (4.66,67) are constructed for the diagonal matrices φ_n. However, the Darboux–Matveev transformations, as well as their iterations are possible for noncommuting matrices too. The determinants $\Delta_n[N]$ (4.67) have the usual definition.

Soliton solutions in a nonlinear theory are distinguished physically – initial perturbations of sufficient amplitude tend to evolve to solitons at long times. In the theory of wave interactions in a state of arbitrary phases the role of solitons is just as important: the stochastization process can be seen as the interaction of an arbitrary background and certain soliton formations that respond to spatial and spectral turbulence [4.23]. The soliton solutions of the discrete Silin–Tikhonchuk equations are obtained if as a starting solution of the system (4.63) the constant

$K_n = \begin{pmatrix} -p & 0 \\ 0 & s \end{pmatrix}$ is chosen. Note that n plays the role of the Langmuir wave frequency.

Now we write the L–A system (4.63) in components of $\Psi_n = \text{diag}\{u_n, v_n\}$

$$\dot{u}_n = -ps u_{n+2}, \quad \dot{v}_n = -ps v_{n+2},$$
$$-p u_{n+1} + u_{n-1} = \mu v_n, \quad s v_{n+1} + v_{n-1} = \lambda u_n. \tag{4.68}$$

The solution of the system (4.68) is found in the form of Fourier integrals

$$u_n = \int_{\Omega_0} f(\omega) \omega^n e^{-ps\omega^2 \Omega} d\omega,$$
$$v_n = \int_{\Omega_0} g(\omega) \omega^n e^{-ps\omega^2 \Omega} d\omega. \tag{4.69}$$

Here ω is an auxiliary integration variable over the interval Ω_0 of the allowed values $\sim k_L$ (Fig. 4.1). Substituting (4.69) into (4.68) we get the condition for the functions $f(\omega)$ and $g(\omega)$:

$$\int_{\Omega_0} \left[\frac{(1 - p\omega^2) f}{\omega} - \mu g \right] \exp(-ps\omega^2 \Omega) d\omega = 0,$$
$$\int_{\Omega_0} \left[\frac{(1 + s\omega^2) g}{\omega} - \lambda f \right] \exp(-ps\omega^2 \Omega) d\omega = 0. \tag{4.70}$$

The condition of solvability of the linear equation system that follows from (4.70) is the biquadratic equation

$$-ps\omega^4 + (s - p - \lambda\mu)\omega^2 + 1 = 0, \tag{4.71}$$

the roots of which, $\pm\omega_1, \pm\omega_2$, define the functions f and g by $|\omega_1| \neq |\omega_2|$:

$$f = c_1^+ \delta(\omega - \omega_1) + c_1^- \delta(\omega + \omega_1) + c_2^+ \delta(\omega - \omega_2) + c_2^- \delta(\omega + \omega_2),$$
$$g = (1 - p\omega_1^2) \left[c_1^+ \delta(\omega - \omega_1) - c_1^- \delta(\omega + \omega_1) \right] / \omega_1 \mu$$
$$+ (1 - p\omega_2^2) \left[c_2^+ \delta(\omega - \omega_2) - c_2^- \delta(\omega + \omega_2) \right] / \omega_2 \mu, \tag{4.72}$$

and by $|\omega_1| = |\omega_2|$,

$$f = A^+ \delta(\omega - \omega_1) + B^+ \delta'(\omega - \omega_1) + A^- \delta(\omega + \omega_1) + B^- \delta'(\omega + \omega_1),$$
$$g = (1 - p\omega_1^2) [A^+ \delta(\omega - \omega_1) + B^+ \delta'(\omega - \omega_1)$$
$$- A^- \delta(\omega + \omega_1) - B^- \delta'(\omega + \omega_1)] / \omega_1 \mu. \tag{4.73}$$

One-soliton solutions are given by the application of the unique transform (4.65)

$$p_n = p \frac{u_{n+1} v_{n-2}}{u_{n-1} v_n},$$
$$s_n = s \frac{u_{n-2} v_{n+1}}{u_n v_{n-1}}.$$

The components of φ_n are found by the substitution of (4.72) into (4.70); ω_i are defined by the values of $\lambda = \lambda_1$, $\mu = \mu_1$ from (4.71)

$$p_n = \frac{\cosh[\Theta_n^- + (1/2)\ln(\omega_1/\omega_2)]\cosh[\Theta_{n-1}^+ + \Theta - (1/2)\ln(\omega_1/\omega_2)]}{\cosh[\Theta_n^- - (1/2)\ln(\omega_1/\omega_2)]\cosh[\Theta_{n-1}^+ + \Theta + (1/2)\ln(\omega_1/\omega_2)]},$$

$$s_n = s\frac{\cosh[\Theta_{n-1}^- - (1/2)\ln(\omega_1/\omega_2)]\cosh[\Theta_n^+ + \Theta + (1/2)\ln(\omega_1/\omega_2)]}{\cosh[\Theta_{n-1}^- + (1/2)\ln(\omega_1/\omega_2)]\cosh[\Theta_n^+ + \Theta - (1/2)\ln(\omega_1/\omega_2)]} \quad (4.74)$$

where

$$\Theta_n^\pm = \frac{ps(\omega_2^2 - \omega_1^2)\Omega}{2} + \frac{n}{2}\ln(\omega_1/\omega_2) + \frac{1}{2}\ln\frac{c_1^+ \pm (-1)^n c_1^-}{c_2^+ \pm (-1)^n c_2^-}$$

$$\Theta = \frac{1}{4}\ln\frac{s\omega_2^2 - p\omega_1^2 + ps + 1}{s\omega_1^2 - p\omega_2^2 + ps + 1}. \quad (4.75)$$

The functions f and g that correspond to multiple roots of (4.71) that are defined by (4.73) give rational solitons after substitution into (4.69):

$$p_n = p\frac{[v_n - q_n^+ + 1/\omega_1][v_{n-1} - q_{n-1}^- - 1/\omega_1]}{[v_n - q_n^+ - 1/\omega_1][v_{n-1} - q_{n-1}^- + 1/\omega_1]},$$

$$s_n = s\frac{[v_{n-1} - q_{n-1}^+ - 1/\omega_1][v_n - q_n^- + 1/\omega_1]}{[v_{n-1} - q_{n-1}^+ + 1/\omega_1][v_n - q_n^- - 1/\omega_1]}, \quad (4.76)$$

where $v_n = n/\omega_1 - 2p\omega_1\Omega$, $q^\pm = (A^+ \mp (-1)^n A^-)/(B^+ \pm (-1)^n B^-)$.
This may also be written

$$p_n = p\frac{z_n^2 - 2.25\omega_1^2}{z_n^2 - 0.25\omega_1^2},$$

where $z_n = (2n-1)/2\omega_1 - 2ps\omega_1\Omega - (A^+ - (-1)^n A^-)(B^+ + (-1)^n B^-)$. The behavior of the functions p_n, s_n [(4.74) or (4.76)] that describe one-soliton solutions of (4.61) demonstrate an interesting property that is absent from other equations for solitons. The values of p_n, s_n as functions of the discrete variable n with $c_{1,2}^- \neq 0$ change by jumps because $c_{1,2}^-$ is multiplied by $(-1)^n$. The physical sense of such "shaking" behavior of the solitons with $\omega \sim n$ is not clear. The assumption of slow variation with n is fulfilled at small $c_{1,2}$ that restricts the amplitude shaking. The classes of these solitons may be of interest to physicists. An interpretation within bifurcation theory may be found. Multi-soliton solutions are obtained if in (4.66,67) the functions $\varphi_n(k)$ defined by (4.69–73) with different $\lambda = \lambda_k$, $\mu = \mu_k$ are substituted. The limit on the values of Ω contains the soliton contributions with phase shifts. If the matrices $b_{mk} = \omega_{mk}^{-2}$, $a_{km} = \ln\omega_{km}$ are introduced, where ω_{mk} are the positive roots of (4.71) with $\lambda = \lambda_m$, $\mu = \mu_m$, then the phase shift of the soliton with number m is equal to

$$\delta_m = \frac{1}{a_{2m} - a_{1m}}\ln\frac{W_3 W_2}{W_4 W_1}$$

where W_k are the Wandermond determinants

$$W_s = W(b_{11}, \ldots, b_{1,m-1}, b_{s,m}, b_{2,m+1}, b_{2N}), \quad s = 1, 2;$$
$$W_r = W(b_{21}, \ldots, b_{2,m-1}, b_{r,m}, b_{1,m+1}, b_{1N}), \quad r = 3, 4.$$

Let us note that (4.61) are invariant with respect to the simultaneous change $\Omega \to -\Omega$, $p_n \to -p_n$, $s_n \to -s_n$. Therefore, in all solutions Ω may be changed to $-\Omega$. Such solutions may be obtained by means of the second L–A pair of (4.61) that can be transformed from (4.63) by the change $K_n \to -K_n$, $\dot{\Psi}_n \to -\Psi_n$. The solutions are also invariant with respect to translation $\Omega \to \Omega - \Omega_0$.

The boundary conditions (4.54) are satisfied by suitable choice of the parameters p, s, $c_k^{\pm}$. It is enough to set $p_n(0) = s_n(0)$. The second boundary (4.55) is due to increasing Landau damping at $\Omega \gg \Omega_m$. In (4.60) or (4.56) for Langmuir waves the relaxation is obviously not taken into account. The relaxation effect may be included via nonsingular perturbation theory as in Sect. 2.4 [see (2.77)]. By including the collisionless damping terms we get the validity of the boundary condition (4.55).

5. Evolution Equations for Internal Waves in Media with Strongly Inhomogeneous Stratification

Here the specific internal hydrodynamical waves that exist in liquids with strongly different scales of stable stratification (Sect. 5.1) are analyzed. The interest in these waves is connected with governing nonlocal evolution equations. The well-known examples of such equations are the Benjamin–Ono and Joseph equations, which are integrable. These equations are generalized for the two-dimensional multimode case in Sects. 5.2,3 for waveguides in the ocean and atmosphere. In Sect. 5.4 it is shown how the growth of the mean free path of particles leads to wave capture and the formation conditions allowing waveguide-type propagation.

5.1 Waves in a Medium with Varying Stratification

We take, as in the previous sections, the case where the parameters of a medium depend only on the variable z. Now let us suppose that the dependence is such that a given range of wavelengths cannot exist in the region $z > h$. Thus, even an infinite medium can become a quasi-waveguide at least for a restricted wavelength range. Atmosphere is an example for this case for all types of waves [5.1–3], and the oceanic thermoclyne for internal and sound waves [5.4,5]. In strongly inhomogeneous plasma there exist surface waves [5.6]. In restricted dielectrics the effect of waveguide propagation arises due to the refraction index change (Sect. 3.4) [5.7]. Under conditions of nonlinear propagation self-focusing may occur. The wave is concentrated around some line in space and an "autowaveguide" appears [5.8].

Due to the localization of the wave in space and in some wavelength range it is natural to use Fourier analysis in the localization interval and to describe the influence of the surrounding medium with appropriate matching conditions at the boundaries. The boundary conditions are determined from the physics of the problem, but primarily from the choice of the basic equations used in the entire area of varying z.

The problems are formulated as before, i.e., we introduce small parameters that characterize the initial conditions: the amplitudes and scales of independent variables. The crux of the task at this stage is the definition of the role of the boundary regime. We adopt the concept of interaction of propagation regions which generally leads to changes in the dispersion law as well as to new values in the nonlinear constants.

We assume that the stratified medium occupies the region $[0, H]$ in z. We separate this interval into two layers $[0, h]$ and $[h, H]$ and assume that quasi-waveguide propagation occurs in the first layer, i.e., a hyperbolic equation holds in some wavelength range. One can imagine three different sets of physical conditions:

(1) The description of both layers is based on a universal set of independent dynamical variables, as in propagation of internal waves in the oceanic thermocline at all depths. Then the equations of an incompressible fluid hold (Sect. 5.2).
(2) The dynamical description changes from layer to layer. This is the case when the atmosphere and ocean or atmosphere and plasma are considered in a hydrodynamical approach. Here the number of dynamical functions may be different for each layer.
(3) The theoretical description varies from layer to layer. For example, in the first layer a hydrodynamical approach is used and in the second layer the kinetic equations (Sect. 5.4).

The matching condition is simplest for the first case. The number and type of necessary conditions do not change at the boundaries when they are approached from different sides. In problems of the second type it is necessary to describe the joint between layers. It is possible to average in the interval to reduce the number of dynamic variables in the vicinity of boundaries. In the third approach it is necessary to use a broader description in the waveguide layer with a restricted distribution function, for example, that of local equilibrium. Previous studies have shown that the most effective matching of wave and nonwave regimes of different layers is obtained by the introduction of an additional transition interval (boundary layer).

In the theory of quasi-waveguide propagation the dispersion and nonlinearity may become nonlocal [5.9–13]. New types of single-mode equations appear, among them those integrable by inverse scattering methods, Benjamin–Ono, Joseph [5.14] and nonlocal Kadomtsev–Petviashvili equations [5.15]. The dispersion that results from nonlocal operators is very similar to KdV dispersion. It is weak at long wavelengths and the operators are invariant with respect to translation [5.16]. The inclusion of mode interaction can be executed by the methods of Sects. 2.4–6, i.e., by nonsingular perturbation theory. Including these effects within the first order perturbation theory allows one to avoid infinite series.

In this section we consider the general solution of the simplest case. We write the initial equations system, the initial and boundary conditions in the standard form

$$\Psi_t + A\Psi = \sigma\nu(\Psi), \quad \Psi = \begin{pmatrix} w \\ \dots \\ \Psi_n \end{pmatrix}, \tag{5.1}$$

$$\Psi\bigg|_{t=0} = \begin{pmatrix} w^0(\varepsilon x, z) \\ \cdots \\ \Psi_n^0(\varepsilon x, z) \end{pmatrix} \quad w \equiv \Psi_1 \, ;$$

$$w(0) = w(H) = 0 \, , \quad \lim_{z \to h+0} w = \lim_{z \to h-0} w \, ,$$

$$\lim_{z \to h+0} w_z = \lim_{z \to h-0} w_z \tag{5.2}$$

The variable w contains the splicing conditions. In hydrodynamics such a quantity may be the projection of the pressure or vertical velocity that describes the mass flow across a surface $Z = h$. In electrodynamics it is usually the tangential component of the electric field and the normal to the magnetic field. The units of measurement and the time interval are chosen so that $|w|$, $|\Psi_k|$, $|\nu(\Psi)| < 1$, $\sigma \ll 1$.

The first step is to describe the waveguide region with relatively weak inhomogeneity. In this region the dispersion relations (2.30) are approximately valid, and the projection operators over the dispersion branches may be introduced. Let us assume that with respect to the small inhomogeneity parameter, the linearized system (5.1) can be approximately reduced to

$$L(w) = 0 \, , \quad L = L_1(x, t) + L_2(x, t) L_z \, . \tag{5.3}$$

Further, let the operator L_z generate a Sturm–Liouville problem after a Fourier transformation which is equivalent to the substitution $w \sim Z(z) \exp i(\varepsilon k x - \omega t)$,

$$L_z Z_n = \lambda_n Z_n \, , \quad Z_n(0) = 0 \, , \tag{5.4}$$

$$Z_n'(h) = \gamma_n(k) Z_n(h) \, . \tag{5.5}$$

The condition (5.5) appears as a result of the smooth splicing of a Fourier wave mode. We notice that the parameter λ_n in (5.4) can enter in a more complex way [5.4,17]. The condition at $z = 0$ also may be more complex. We assume for simplicity that L_z is a differential operator of second order. The results may be generalized in a straightforward manner.

The method that accounts for weak dispersion of the mode functions Θ_μ generated by the layer $[h, H]$ is the following. The function w in the interval $[0, h]$ in the zeroth approximation is represented as an expansion in the basis functions Z_n that determine Θ_n:

$$w = w_{(0)} = \sum_k \Theta_k Z_k \, . \tag{5.6}$$

The identity

$$L_1(Z_n, w) + L_2(Z_n, L_z w) = L_1(Z_n, w) + L_2(L_z^+ Z_n, w)$$
$$+ L_2 \Gamma_n w \bigg|_{z=h} = 0 \, , \tag{5.7}$$

$$\Gamma_n w \bigg|_{z=h-0} = Z_n(h) \lim_{z \to h+0} (\tilde{w}_z - \gamma_n \tilde{w}) \, , \tag{5.8}$$

where $\tilde{w}$ is the solution for the layer $[h, H]$, holds. Equation (5.7) results as the scalar product of (5.3) and Z_n. The latter term in (5.7) appears after integration by parts as a result of operating with L_z on the left argument of the scalar product. Equality (5.8) is a direct consequence of the continuity of the function w and its derivative (5.2).

Let the function $\tilde{w}$ in the interval $z \in [h, H]$ satisfy

$$L_0 \tilde{w} = 0 . \tag{5.9}$$

The boundary condition at $z = H$ follows from (5.2):

$$\tilde{w}\Big|_{z=H} = 0 . \tag{5.10}$$

At the boundary $z = h$, in the neighborhood of which the properties of the initial system are changed (the operator L is reduced to L_0), the boundary regime is given by (5.2)

$$\lim_{z \to h+0} \tilde{w} = \lim_{z \to h-0} w , \quad \lim_{z \to h+0} \tilde{w}_z = \lim_{z \to h-0} w_z . \tag{5.11}$$

It is namely in these equalities that the iteration procedure is realized. To first order in the right side of (5.11) the zeroth approximation in ε is substituted

$$w = \sum_{j=0}^{\infty} \varepsilon^j w_{(j)} ,$$

where $w_{(0)}$ is represented in the form (5.6). The first step in the definition of $\tilde{w}$ is therefore

$$\tilde{w}\Big|_{z=h} = \sum_n Z_{n(0)}(h) \Theta_n(x,t) , \tag{5.12}$$

where $Z_{n(0)}$ is determined by (5.4) and the boundary condition (5.5) at $\varepsilon = 0$.

Effective application of this method is possible if the solution of (5.9,10,12) allows the explicit representation that gives $\Gamma_n w \Big|_{z=h}$, for example, as a sum of the results of mode functions operated on by linear operators:

$$\Gamma_n w \Big|_{z=h} = \varepsilon \Gamma_n w_{(1)} \Big|_{z=h} = \varepsilon \Gamma_n D w_{(0)} = \varepsilon \sum_m \partial_{mn} \Theta_m . \tag{5.13}$$

The smooth matching of the function w at the boundary between regions of different stratification closes the equation system for Θ_n (5.9). The equation for the next order perturbation theory can easily be constructed:

$$L_1(Z_n, w_{(1)}) + \lambda_n L_2(Z_n, w_{(1)}) + \varepsilon L_2 \Gamma_n w_{(2)} \Big|_{z=h} = 0 ,$$

$$\varepsilon \Gamma_n w_{(2)} \Big|_{z=h} = \varepsilon Z_n(h) \lim_{z \to h} D w_{(1)} ,$$

where $Dw_{(1)}$ is a solution (5.9,10) when $\tilde{w} = w_{(1)}\big|_{z=h}$. All the quantities should be calculated to first order in ε.

The idea of the mode decomposition is useful if it is restricted to a small number of finite amplitude modes [5.17,18]. In the laboratory it can be obtained by special excitation conditions, and in geophysical systems a similar situation can be observed. Usually it is assumed that the effects of nonlinear dispersion, together with dissipation, filter higher or short wavelength modes. Therefore, equations such as the Benjamin–Ono are themselves of physical interest [5.19,20]. For the derivation of such equations it is enough to make one iteration in the sums (5.6,12,13). As was noted above, the weak mode interaction in the long wavelength range may be taken into account by iterations in the amplitude σ.

It is clear that if nonlinearity exists in the basic equations, it must be accounted for along with all dispersion effects at every stage of the perturbation theory treatment. This method is successful if it can be limited to the lowest order modes. The contribution of nonlinear terms due to interaction between layers can be considerable. It is particularly strong if the mode amplitude grows as the boundary is approached, as in the case of atmospheric internal waves (Sects. 2.3,5.3,4). For the derivation of nonlinear model equations the equation of coupling between components Ψ_i is necessary: $\Psi_i = S_i w$. For any component we can write the expansion to the same order of ε, $\Psi_i = \sum_n \Psi_{in} \tilde{Z}_n$, and using either the orthogonality conditions for $Z_k(\tilde{Z}_k)$ or the eigenfunctions of the conjugate equation we project the equations of the system (5.1) onto the mode basis as in Sects. 2.2,3 or 3.1. Then we obtain a system of the type (2.50)

$$\Psi_{nt} - A_n \Psi_n = \sigma \nu^n(\Psi) , \qquad (5.14)$$

where the operators of interaction between layers ∂_{mn} are included in A_n. It can happen that the act of projection changes the magnitude of the coefficients. Then it is necessary to renormalize the variables and the parameters of $P_n^{(i)}$ (Sect 6.3). The linearized equation (5.14) determines the projecting operators $P_n^{(i)}$ of the i-dispersion branch of the mode n in the quasi-waveguide. The initial conditions for a projection are given by $\Psi_n^{(i)}\big|_{t=0} = P_n^{(i)} \Psi_n(t=0)$.

We now turn our attention to the definition of the basis functions. For long wavelength waves where the dispersion function $\omega(k)$ in the neighborhood of $k = 0$ is linear, the coefficients $\gamma_{n0} = \gamma_n$ ($\varepsilon = 0$) do not depend on k, e.g., as in surface waves on water, internal gravitational waves, or ion-acoustic plasma waves.

For other dispersion branches we have nonlinear dependence in $\omega_n(k)$. In this case it is possible to use the localization of an initial wave train in a narrow range of k and apply the apparatus of slowly varying amplitudes. In the next paragraphs we shall show how the evolution equations are derived for waves in quasi-waveguides. We shall demonstrate how the theory is generalized to the more complicated second and third approaches listed above. It is clear that in this approach one can consider more than two intervals (Sect. 5.2). If necessary, one can introduce a transition layer in the absence of a discontinuity in the properties of the perturbed medium. The width of the transition layer can be kept small in relation to the transverse dimensions of the waveguide and thus be approximated as a small parameter. It is also possible that the physical picture demands the choice of the matching boundary h. Thus the dependence on $Z_n(k)$ is in the long wavelength approximation and the dependence on k of the coefficients of the operators is changed (Sect. 5.4).

5.2 Analogues of the KdV and KP Equations with Nonlocal Dispersion

Many works are devoted to the theory of single-mode internal waves with nonlocal dispersion in the oceanic thermoclyne. *Philips* and *Whitham* considered a narrow thermoclyne in their derivation of the dispersion relation [5.21]. Later *Benjamin* and *Ono* derived the first model equation which allows a dynamical description [5.10,11]. These authors found soliton solutions of the equations. It was assumed that the thermoclyne is surrounded by infinite layers. Finite intervals were studied by *Joseph* [5.13]. The method of successive derivation of the evolution equations on the basis of hydrothermodynamics with different ground states is contained in [5.4,22], in [5.22] flow is taken into account. The generalization to cases of wave fronts with different geometry is given in [5.23]. A discussion of boundary matching conditions and an investigation of mode interactions in three dimensions is given in [5.9,17]. Below we shall reproduce the main steps of the derivation with particular attention paid to boundary conditions [5.9]. In addition, we shall calculate the constants in the matching conditions.

The integrability of the Benjamin–Ono and Joseph equations is stated in [5.14,15]. In [5.15] the two-dimensional generalization was studied. Many-soliton [5.14,24–26] and finite-gap solutions [5.27] have been constructed. Here we also give the explicit form of the soliton solutions of the equations.

The equations of hydrodynamics that are valid for internal water waves are discussed in detail in Sect. 2.1. Here we reproduce the formulas (2.15,16), neglecting the viscosity and thermoconductivity and use the Boussinesq approximation [5.4]. The essence and validity of these approximations for long wavelength waves has been discussed [5.4]. Let us write, for simplicity, the vector equation of motion in components. As before, we shall denote the velocity vector components $v_1 = v_x = u; \ldots$ The incompressibility condition in the absence of thermoconductivity (Sect. 6.2) gives

$$\text{div } \boldsymbol{v} = 0 \ . \tag{5.15}$$

The remaining equations are

$$\begin{aligned}
\varrho_0 u_t &= -\varrho(\boldsymbol{v}, \nabla)u = p_x \ , \\
\varrho_0 v_t &= -\varrho_0(\boldsymbol{v}, \nabla)v - p_y \ , \\
\varrho_0 w_t &= -\varrho_0(\boldsymbol{v}, \nabla)w - p_z - \varrho' g \ , \\
T'_t - w\bar{T}_z &= -(\boldsymbol{v}, \nabla)T' \ .
\end{aligned} \tag{5.16}$$

The state equations coincide with (2.10) and have already been taken into account.

The derivation of model equations could be achieved with the use of projection operators on the subspaces of the dispersion branches. However, in this problem this leads to superfluous terms. It is well known that the wave perturbations in (5.15,16) contain only the contribution of two branches, i.e., the left and right internal waves, since the acoustic waves are excluded by the incompressibility condition and the Rossby waves by the zero Coriolis parameter. These branches are effectively split by a two-parameter decomposition [5.18]. The choice of the initial condition is successfully realized by the range of the wave vector and the dispersion relation, since in a fluid, due to the low compressibility, dispersion branches are well separated. Therefore, we shall transform, as in Sect. 5.1, the system (5.15,16) to a single equation in first order of the amplitude. Differentiating by the coordinates we combine the equations (see details in [5.17]) and get for the vertical velocity projection w

$$\Delta w_{tt} + \alpha g \bar{T}_z \Delta_n w = \text{div} \left\{ [(\boldsymbol{v}\nabla)\boldsymbol{v}]_z - \nabla(\boldsymbol{v}\nabla)w \right\}_t - g\alpha \Delta_h (\boldsymbol{v}\nabla)T' , \quad (5.17)$$

where $\Delta_h w = w_{xx} + w_{yy}$.

Let us assume that in the region of waveguide propagation (in a thermoclyne) the Väsälä frequency N varies weakly with z. Then by means of the local dispersion relation

$$k_z^2 \omega^2 = (N^2 - \omega^2)(k_x^2 + k_y^2) \quad (5.18)$$

one can estimate the scales k_x and k_y for the structure of the given vertical mode and the frequency range. For long period waves $\omega \ll N$. It follows from (5.18) that

$$\lambda_z^2 \ll \frac{\lambda_x^2 \lambda_y^2}{\lambda_x^2 + \lambda_y^2} , \quad \lambda_z \approx \frac{\omega \lambda_x}{\bar{N}} , \quad (5.19)$$

where $\bar{N}$ is the average Väsälä frequency in the thermoclyne. In correspondence with the traditional assumption about the two-dimensional instability of KdV solutions (Sect. 2.4) we shall assume that $\lambda_x \ll \lambda_y$. Such disturbances have a larger scale along the wave front than a longitudinal one. From (5.19) it follows immediately that $\lambda_z \ll \lambda_x$. All these assumptions lead to a set of restricted initial conditions. We introduce the small parameters $\beta = \lambda_z/\lambda_x$, $\nu = \lambda_z/\lambda_y$. We exclude the functions u, T' from the small terms and obtain the linear relations as in Sect. 2.4:

$$u = -\int_{-\infty}^{x} w_z dx , \quad T' = -\bar{T}_z \int_0^t w dt . \quad (5.20)$$

We suppose that at times $t \leq 0$ and at $x \to -\infty$ the perturbation is absent. Going to the dimensionless variables $x_i = \lambda_i x'_i$, $t = 2\pi t'/\bar{N}\beta$, with the aid of the other linear relations we write $u = \lambda_z \bar{N} u'/2\pi$, $v = \nu \lambda_z \bar{N} v'/2\pi$, $w = \beta \lambda_z \bar{N} w'/2\pi$, $\alpha T' = T$, $N = \bar{N} N'$. It can easily be seen that the amplitude of v_y

in the divergence of velocity is small and should not be included in (5.20). The integrals over x can be removed if the current function Ψ is introduced so that $w' = \Psi_x$, $u' = \Psi_z$. This function allows the closed equation (4.17) to be more compact (the prime on the dimensionless Väsälä frequency is omitted):

$$\Psi_{zztt} + N^2\Psi_{xx} = -\beta^2\Psi_{ttxx} - \nu^2 N^2\Psi_{yy} + (\Psi_z\Psi_{zx} - \Psi_x\Psi_{zz})_{zt}$$
$$- \left[N^2\Psi_x\int_0^t \Psi_{xx}dt - \Psi_x\left(N^2\int_0^t \Psi_x dt\right)_z\right]_x. \quad (5.21)$$

The linear operator in (5.21) coincides with the analogous operators in (5.17) and (5.3) when $L_z f = -f_{zz}$ and $L_2 f = -f_{tt}$. The operator L_1 is given by $L_1 f = N^2 f_{xx} + \beta^2 f_{xxtt} + \nu^2 f_{yy}$ and determines the small dispersion $\sim \beta^2, \nu^2$. Because the main part of (5.21) contains N^2 the separation of variables gives the equation for the basis functions

$$-Z_{zz} = \frac{N^2}{c^2}Z, \quad (5.22)$$

where the linear propagation velocity c is the eigenvalue.

The problem is solved in the following coordinate system. The origin is situated in the middle of the thermoclyne which is the waveguide layer occupying the interval $[-h, h]$. The surface of the ocean has the coordinate $z = H_1$ and the bottom has $z = -H_2$. The z axis is directed upward.

First we attempt the solution outside the thermoclyne. On the ocean surface we set the condition of surface wave filtration and on the ocean bottom the condition of impenetrability:

$$w\bigg|_{H_1} = w\bigg|_{-H_2} = 0. \quad (5.23)$$

The solution for the upper layer that is adjacent to the ocean surface will be denoted by the index + and the solution for the bottom layer by the index −. We have, due to (5.23) and the definition of Ψ,

$$\Psi^+(x, H_1) = 0, \quad \Psi^-(x, -H_2) = 0.$$

Outside of the thermoclyne the gradients of the background temperature $\bar{T}_z$ and therefore the Väsälä frequency (N) are small. The amplitudes also exponentially decrease, because of $\text{Re}\{k_z\} = 0$, with increased distance from the thremoclyne. Taking into account these assumptions and the absence of disturbances at $t \leq 0$ it is assumed that

$$L_0\Psi = \Delta\Psi = 0,$$

[compare with (5.9)]. When $\lambda_z \ll \lambda_y$, which is valid also outside of the thermoclyne, the y derivatives can be ignored. Finally,

$$L_0\Psi^\pm = \Psi^\pm_{xx} + \Psi^\pm_{zz} = 0. \quad (5.24)$$

Thus we have two Dirichlet problems for the layers $z \in [h, H_1], [-h, -H_2]$ with zero valued functions at the exterior and with $\Psi^{\pm}$ at the interior boundaries.

The solution of the standard Dirichlet problem for the interval $\tilde{z} \in [0, \pi]$ is expressed by an integral over boundaries 9vccca[5.28]. Because at $z = \pi, \Psi^0 = 0$ the solution has the form

$$\Psi^0(x, \tilde{z}) = \frac{1}{\pi} \int_{-\infty}^{\infty} d\xi \Omega(\xi) \frac{\sin \tilde{z} \exp(x - \xi)}{1 - 2\exp(x - \xi)\cos \tilde{z} + \exp 2(x - \xi)} \; . \quad (5.25)$$

Transforming (5.25) as in [5.9] and differentiating it as a generalized function, we get its limit at $z \to 0$

$$\Psi_z^0(x, 0) = \frac{\partial \tilde{z}}{\partial z} \frac{\partial}{\partial x} \int_{-\infty}^{\infty} \Omega(x - \xi) \coth \frac{\xi}{2} d\xi \; . \quad (5.26)$$

The function $\tilde{Z}(z)$ transforms the layer $[0, \pi]$ into the layer $[-h, -H_2]$ or $[h, H_1]$. Here $\tilde{Z}^- = \pi(z + h)/(h - H_2)$, $\tilde{Z}^+ = \pi(z - h)/(H_1 - h)$ are the corresponding notations for the upper and lower layers. Thus, by (5.26)

$$\Psi_z^+(x, h) = \frac{1}{2(h - H_1)} \int_{-\infty}^{\infty} \coth \frac{\xi}{2} \Psi_x(x - \xi, h) d\xi \; ,$$

$$\Psi_z^-(x, -h) = \frac{1}{2(H_2 - h)} \int_{-\infty}^{\infty} \coth \frac{\xi}{2} \Psi_x(x - \xi, -h) d\xi \; . \quad (5.27)$$

Now we put instead of Ψ the single-mode Fourier component $\Psi = Z(z) \times \exp i(kx - \omega t)$ and find, as is shown in Sect. 5.1, for smooth matching

$$Z_z(h) = \frac{ik}{2(h - H_1)} \int_{-\infty}^{\infty} \coth \frac{\xi}{2} e^{-ik\xi} d\xi Z(h) \; ,$$

$$Z_z(-h) = \frac{ik}{2(H_2 - h)} \int_{-\infty}^{\infty} \coth \frac{\xi}{2} e^{-ik\xi} d\xi Z(-h) \; . \quad (5.28)$$

The integrals in (5.25-28) are the principal values, while (5.28) is a Fourier transformation of a general function. It is important to note that the coefficients that relate the value of the function Z and its derivative at the boundary do not depend on the eigenvalue c (or the mode number). This allows one to obtain the orthogonality of eigen functions in the usual manner. Tables of integrals give $\int_{-\infty}^{\infty} e^{-ik\xi} \coth \frac{\xi}{2} d\xi = \frac{2\pi}{i} \coth \pi k$. In the long wavelength limit $\frac{2\pi}{i} \coth \pi k \to 2/ik$. Therefore,

$$Z'(h) = \frac{Z(h)}{h - H_1} \; , \quad Z'(-h) = \frac{Z(h)}{H_2 - h} \; . \quad (5.29)$$

Thus, for the upper and lower edges of the thermoclyne the third kind of conditions with $\gamma_0^+ = 1/4\pi(H_1 - h)$, $\gamma_0^- = 1/4\pi(h - H_2)$ are obtained (5.5). Equation (5.4) together with the boundary conditions (5.29) form the Sturm-Liouville problem that determines the orthogonal basis functions Z_n. Expansion in this basis gives the mode representation for whose coefficients we intend to derive the evolution equations. The orthogonality condition $\int_{-h}^{h} Z_m Z_n N^2 dz = (Z_m, Z_n) = 0$ simplifies the form of these equations. Equation (5.7) in this case of two matching boundaries contains two operators of the "interaction across the boundary" $\Gamma_h^{\pm} = \partial/\partial z + \gamma_0^{\pm}$ and has the form

$$\int_{-h}^{h} Z_n \Psi_{zztt} + \int_{-h}^{h} N^2 \Psi_{xx} Z_n dz$$

$$= \Theta_{xx}^n - \frac{\Theta_{tt}^n}{c_n^2} + \frac{\partial}{\partial t^2}\left[Z_n(h)(\Psi_z^+ + \gamma_0^+ \Psi^+)\right]\bigg|_h$$

$$- Z_n(-h)(\Psi_z^- + \gamma_0^- \Psi^-)\bigg|_{-h}\bigg] .$$

Let us put in the boundary the functions of the next approximation that are defined by a smooth matching condition, i.e., (5.27) through the boundary values of the zeroth approximation $\Psi(x, \pm h) = \sum_n Z^n(\pm h) \Theta^n(x, y, t)$:

$$\Theta_{xx}^n - \frac{\Theta_{tt}^n}{c_n^2} + \frac{1}{2}\sum_k \frac{\partial^2}{\partial t^2}\left[\frac{Z_n(h)Z_k(h)}{h - H_1} - \frac{Z_n(-h)Z_k(-h)}{H_2 - h}\right]$$

$$\times \left[\int_{-\infty}^{\infty} \coth\frac{\xi}{2}\Theta_x^k(x - \xi)d\xi + 2\Theta^k\right]$$

$$= -\beta^2 \sum_k \Theta_{xxtt}^k \int_{-h}^{h} Z_m Z_n dz - \nu^2 \Theta_{yy}^n$$

$$+ \sigma \sum_{m,k}\Bigg\{(\Theta^m \Theta_x^k)_t \int_{-h}^{h}\left(Z_z^m Z_z^k + Z^m \frac{N^2}{c_k^2} Z^k\right) Z^n dz$$

$$- \left[\Theta^m \int_0^t \Theta_{xx}^k dt \int_{-h}^{h} Z^m Z^n Z^k N^2 dz - \Theta_x^n \int_0^t \Theta_x^k dt\right.$$

$$\times \left.\int_{-h}^{h} Z^m(N^2 Z^k)_z Z^n dz\right]_x\Bigg\} . \tag{5.30}$$

The right-hand side of (5.30) contains the obviously small parameters whose values are given by the choice of the initial conditions. The boundary nonlocal operator, when H_i is larger than h also has a small coefficient because the normalized functions $Z_n(\pm h)$ are of the order of unity. Note that if the boundaries $H_{1,2}$ are removed far from the thermoclyne the system may be considered as a Dirichlet problem for equation (5.24) at a half plane. Then, instead of the Joseph operator, we have the Benjamin–Ono operator in (5.30) that is defined via a Hilbert transformation H: $V_{BO} = H\partial/\partial x$ [5.17].

Equation (5.30) describes waves that propagate in both directions along x. The oppositely directed localized waves that have a large relative velocity interact for a short time. Thus coupling effects maybe neglected if initially a monodirectional wave is excited. Therefore we consider separately mode systems for every direction that interact between themselves. Introducing $\xi = x - c_n t$, $\tau = \sigma t$ and taking σ as the maximum of the small parameters we integrate (5.30) in new variables, neglecting terms of order σ^2. After division by $2c_n$ and returning to variables x, t we have

$$\Theta_t^n + c_n \Theta_x^n + \sum_{m,k} g_{mkn} \Theta^m \Theta_x^k + \beta^2 \sum_m d_{nm} \Theta_{xxx}^m + \nu^2 \frac{c_n}{2} \int_{-\infty}^x \Theta_{yy}^n d\xi$$

$$+ \sum_m d_{nm}^1 \int_{-\infty}^\infty \left\{ \left[\coth \frac{\xi}{2} + \mathrm{sgn}(x - \xi) \right] \Theta^m(x\quad \xi) \right\}_{xx} d\xi = 0 , \quad (5.31)$$

where

$$g_{mkn} = \frac{c_n^2}{2} \int_{-h}^h Z_n \left[\left(\frac{1}{c_n^2} - \frac{1}{c_k^2} \right) (N^2 Z^k)_z Z^m \right.$$

$$\left. + N^2 \left(Z_z^m Z^k \frac{1}{c_n^2} + Z^m Z_z^k \frac{1}{c_m^2} \right) \right] dz , \quad (5.32)$$

$$d_{nm} = c_n^3 \int_{-h}^h \frac{Z_m Z_n dz}{2} , \quad (5.33)$$

$$d_{nm}^1 = \frac{c_n^3}{2} \left[\frac{Z_n(h) Z_m(h)}{h - H_1} - \frac{Z_n(-h) Z_m(-h)}{H_2 - h} \right] . \quad (5.34)$$

At a constant Väsäla frequency the form of the nonlinear and first dispersion coefficients is simplified to

$$g_{mkn} = \frac{c_n^2 N^2}{2} \int_{-h}^h Z_n \left[\left(\frac{1}{c_n^2} - \frac{1}{c_k^2} + \frac{1}{c_m^2} \right) Z^m Z_z^k + \frac{1}{c_n^2} Z^k Z_z^m \right] dz ,$$

$$d_{nm} = \frac{c_n^3 \delta_{mn}}{2N^2} .$$

We note moreover that in the general case the dispersion is described by both KdV and nonlocal contributions. It can be said that if the ocean boundaries tend towards the thermoclyne boundaries, then only KdV dispersion is retained. Single-mode equations can be easily written if the intermode interaction is neglected. Equation (5.31) generalizes KdV, KP, BO and Joseph equations for the multi-mode situation. Each of the cited equations can be obtained if all parameters except for one tend to zero. Therefore, the solutions of these equations can be used in multi-mode perturbation theories in Sects. 2.5,6. Expressing Θ^n as in (2.87) we obtain (2.88) with another dispersion operator that is defined by (5.31–34).

We now discuss the properties of the solutions of the KdV, BO and Joseph equations [5.1,7,21,29]. Let all the solitons have the same amplitude and the equations have the same nonlinear term. The BO and KdV equations are the results of (5.31) in the sense mentioned above:

$$\Theta_{\text{KdV}} = \frac{a}{\cosh^2(x/b)}, \quad b = \sqrt{\frac{6}{a}}, \quad c = \frac{2a}{3}, \tag{5.35}$$

$$\Theta_J = \frac{2a}{\cosh(x/b) + \cos x}, \quad b = \frac{3}{2\pi a},$$

$$c = \frac{3}{2\pi^2} - \frac{2\arctan 2\pi^2 a}{3} \approx \frac{4\pi^2 a^2}{9}, (a \ll 1), \tag{5.36}$$

$$\Theta_{\text{BO}} = \frac{a}{1 + x^2/b^2}, \quad b = \frac{2}{a}, \quad c = \frac{a}{2}. \tag{5.37}$$

In the formula for a Joseph soliton (5.36) the values of parameters are chosen so that the KdV can be obtained as a limiting case. It can be seen that the Joseph soliton also has intermediate parameter values.

It should be noted that in comparing soliton parameters, i.e., their velocities and form, with parameters of real solitary waves produced in a laboratory the authors of [5.19] showed the preference of a KdV soliton. The authors considered formulas such as (5.35-37) as possible descriptions of solitary waves. The analysis showed that since in reality all vertical dimensions as well as horizontal dimensions are comparable, both dispersion contributions (Joseph and KdV) should be taken into account, as follows from (5.31).

5.3 Quasi-Waveguide Propagation of Internal Waves in the Atmosphere

The structure of the undisturbed or averaged state of the atmosphere is such that the dependence of the gas temperature $\bar{T}(z)$ on the height oscillates. At low altitudes the temperature falls, then after reaching a minimum at about 15 km (tropopause) it begins to increase and reaches a maximum at ≈ 40 km (stratopause). It then decreases again up to the mesopause ($\simeq 80$ km). Finally it tends to a constant value in the thermosphere (> 80 km). The location of the extrema and the amplitude of the oscillations depend on the solar activity and season. The variation of temperature with height is rather significant, reaching hundreds of degrees, and influencing the propagation of internal waves [5.2].

In Chap. 2, internal waves in an exponential atmosphere were considered. It is clear that such an approximation, corresponding to a mean homogeneous temperature $\bar{T}$, can hold only if the height interval is bounded. It is possible, therefore, either to construct a multi-layer atmosphere or to neglect the attractive exponentiality, and to solve the problem of finding the basis functions for realistic approximate $\bar{T}(z)$ or $H(z)$ (Sect. 2.4). We suppose that, as in the ocean thermoclyne, in a given height and frequency interval there exist internal waves (IW). Their existence depends on the value of the critical Väsälä frequency which in turn is determined by the height scale H: $\omega_g = \sqrt{[(\gamma - 1)g]/\gamma H}$. This conclusion is made by the analysis of the dispersion relation that is obtained when one assumes that the vertical wavelength of the IW is smaller than the inhomogeneity scale of the atmosphere. Therefore, the result is approximate. Despite this we will consider the two layer atmosphere model in a region where these assumptions are valid, with constant parameters $H = H_{1,2}$. We assume that the temperature which is proportional to H in the upper layer is higher than in the lower layer. Examples of such a situation are the splitting of the troposphere–mesosphere by the tropopause, or the thermosphere–mesosphere system. The latter is situated at a height of 80 km and is stronger, the temperature change reaches 100 K, or 30 %.

Let us write the dispersion relation given by (2.56) in dimensional form at $a \to \infty$, $\Omega = 0$ but taking into account the vertical acceleration

$$\omega^2(\omega^2 - \omega_a^2) - c^2\omega^2 k^2 + c^2\omega_g^2 k_\perp^2 = 0 , \tag{5.38}$$

where $\omega_a = \gamma g/2c$, $\omega_g = \sqrt{\gamma - 1} g/c$ are the limiting frequencies of the acoustic and gravitational spectra and $c = \sqrt{gH\gamma}$ is the sound velocity; $\gamma = c_p/c_v$. The sound velocity, and therefore the limiting values of $\omega_{a,g}$, depend on H, γ which in turn depends on the altitude. The separation of the acoustic and internal waves is accomplished by the projection procedure described in Appendix 3.

Figure 5.1 shows the dispersion curves for the IW that restrict the regions of possible $\omega, k_\perp$ at a given k_z, H. The index i (ω_{gi}) indicates the atmosphere layer. The higher the temperature of the upper layer, the lower its boundary curve lies. If the internal wave is excited in the layer closest to the earth and its parameters lie in the cross-hatched region of the figure, it cannot generate oscillations in the upper waveguide. The values of $\omega, k_\perp$ do not come under the curve ω_{g2}. The perturbation of the dynamical variables exponentially decreases with height up to the layer boundary since k_z is a pure imaginary number.

We need to find the solution of the equation for the upper layer corresponding to (5.38) with $H = H_2$ or at such values of the wave parameters $\omega, k_\perp$ which give Im $\{k_z\} \neq 0$. The statement of the problem for the upper layer conforms to the formulations of Sect. 5.1. At infinity ($z \to \infty$) we set $w^+ \to 0$ and at the matching boundary we specify the regime on a given frequency–scale range (from the cross-hatched region). Namely, we shall require continuity in the vertical flux given by w and w_z and which follows from the pressure continuity [5.2]. We consider the two-dimensional problem and suppose that in a region 1 the wavetrain with carrier frequency ω_0 and horizontal wavevector $k_x = k_0$ propagates. In region 2 let the condition

$$\omega_0^2(\omega_0^2 - \omega_{a2}^2) + c^2(\omega_{g2}^2 - \omega_0^2)k_0^2 = c^2\omega_0^2 k_z^2 < 0 , \tag{5.39}$$

be fulfilled and the parameters ω_o, k_0 satisfy (5.38) with $\omega_a = \omega_{a1}$, $\omega_g = \omega_{g1}$. Then, taking into account that the upper layer is excited at an almost fixed frequency, we shall state that the vertical projection of the hydrodynamic velocity in this layer satisfies the Helmholtz equation given by (5.39) or by the system of initial equations for this atmosphere given in Sect. 2.1:

$$\left\{ \frac{\partial^2}{\partial z^2} + \frac{\omega_0^2 - \omega_{g2}^2}{\omega_0^2} \frac{\partial^2}{\partial x^2} + \frac{\omega_0^2 - \omega_a^2}{c^2} \right\} w^+ = 0 . \tag{5.40}$$

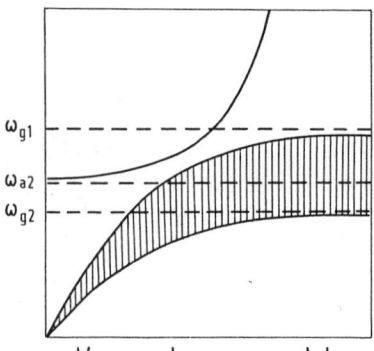

Fig. 5.1. Dispersion curves for internal gravity waves in the atmosphere of two adjacent layers with height scales $H_{1,2}$. Atmospheric quasi-waveguide propagation is obtained for ω, k in the cross-hatched region

Obviously for $\omega_0 > \omega_{g2}$ the coefficient of $\partial/\partial x^2$ is positive and after a scale transformation $x' = x\omega_0/\sqrt{\omega_0^2 - \omega_{g2}^2}$, (5.40) is reduced to the standard form

$$\left(\Delta + \frac{\omega_0^2 - \omega_a^2}{c^2}\right) w^+ = 0 \, . \tag{5.41}$$

It is necessary to construct a solution of (5.41) with a given value at the boundary $z = h$. Such a Dirichlet problem is solved in Sect. 3.4; see (3.65,66). The solution given there differs only by a constant at w^+: $\varepsilon/c^2 \to (\omega_0^2 - \omega_a^2)/c^2$ and by a constant in the boundary condition $Z'(h) = -S(k_0)Z(h)$, where now

$$S(k_0) = \sqrt{c^{-2}(\omega_0^2 - \omega_{a2}^2) + k_0^2(\omega_{g2}^2 - \omega_0^2)/\omega_0^2} \, . \tag{5.42}$$

The solution of the internal wave propagation problem with the upper boundary condition of the third kind is given in Sect. 5.4. The model equation that determines the evolution of the IW wavepacket with a narrow spectral width therefore coincides with (3.67) if the contribution of the lower boundary is neglected and the constants are changed according to (5.41,42). The dispersion operator has the same form as in (3.67). We emphasize that the propagation of this internal wave is not rigorously waveguide–like, since the spectral content varies in the process of evolution. Harmonics appear that do not belong to the cross-hatched region of Fig. 5.1 and which penetrate the boundary. Similar results are reported in [5.30].

5.4 Waves in Gases Inhomogeneous in Knudsen Number

The kinetic equations can, in principle, provide a good description of wave processes in gases. However, the Boltzmann equation and its generalizations are complex and thus progress in finding solutions is slow [5.31–34]. Moreover, the distribution function provides too much information which cannot be measured in a laboratory experiment. A more effective wave description can be based on a simplified approach that follows from a restricted form of the distribution function. A classic example of this is the local equilibrium distribution function [5.31,32]. The corresponding hydrodynamical approach is based on the assumption that the molecular mean free path l is on the order of the wavelength λ. We introduce the Knudsen number $Kn = l/\lambda$. The criterion for the validity of the hydrodynamic approximation is usually the smallness of the Knudsen number, $Kn \ll 1$, or the smallness of the ratio of the process frequency to the collision frequency. Violation of these conditions requires the kinetic equations to be applied. It may be possible, however, to simplify them by using other small parameters which depend on the manner in which the problem is stated. Thus, in the solution to sound propagation problems a number of questions arise concerning the region of a steep wavefront. In this region field gradients are large and therefore the scales of λ are small. A successful description using the kinetic approach to the Knudsen regime has been obtained [5.35]. In problems of vapor condensation a series of works has been devoted to a procedure by which Knudsen and hydrodynamical regimes are matched by the condition of flux continuity. The parameters of condensation kinetics are found, which allows one to get satisfactory agreement with experimental data [5.31,33].

A gaseous atmosphere in a gravitational field is a good example of a medium with inhomogeneous Knudsen number. The decrease in the gas density with increasing height automatically leads to a decrease in the collision number and an increase in the molecular mean free path. Since the mean free path is $l \sim 1/n\sigma_0$, where n is the molecular concentration and σ_0 is the scattering cross-section, the exponential decrease in the density yields an increase in the mean free path that is also exponential. This increase is rapid: at altitudes typical for atmospheric waves, 0–100 km [5.2], the path length varies over 9 orders of magnitude [5.36] and reaches a value of hundreds of kilometers. One can conclude that any atmospheric wave that can propagate vertically, finally reaches the level at which the approximation of hydrodynamic flow fails. Therefore it follows that even the hydrodynamical approximation in the lower layers requires revision as the statement of the problem for it appears to be incomplete.

It is necessary to find a more precise definition of the Knudsen regime for waves in inhomogeneous media. If a molecule is placed at some point in the medium with a nonzero density gradient, its mean free path length will depend on the direction of flight. Anisotropy will be most important in the region where the path length and the scale of the inhomogeneity are similar. If the scale of the inhomogeneity is large, as it is in the atmosphere, then the length of the mean free path may vary strongly depending on the direction even in the order of magnitude. A limiting case is a discrete boundary where the mean free path length also changes discretely. It is clear that the mean free path is larger on the side of decreasing than on the side of increasing density. The local mean free path is determined now by the molecular density n, the scattering cross-section σ_0 and the density gradient. The latter makes the free path length a vector quantity $l_i = l_0 - l_1 \partial n/\partial x_i$.

Internal waves in the atmosphere (Sect. 2.4) represent a non-one-dimensional process whose scales differ in the vertical and horizontal directions. Therefore, the fundamental inhomogeneity of the atmosphere that induces the vector properties of the free path length should be compared with the vector parameter of a wave: with a wave vector, for example. Thus, the role of the Knudsen number must be taken by the tensor

$$K_{is} = (l_0 - l_1 n_{x_i}) k_s \ . \tag{5.43}$$

Now each direction is characterized by a given ratio of the path length to the inhomogeneity scale along this direction. The question of the validity of hydrodynamical approximation under these circumstances requires an analysis which is much more complicated than for homogeneous media. The kinetic equation will contain the group of parameters that are determined by the tensor components (5.43). We shall estimate the diagonal components and choose the Knudsen regime on this basis. It is clear at least that one should compare the horizontal and vertical scales of the wave with the horizontal and vertical mean path lengths, respectively.

In this section the two-dimensional model is described. It is the basis for the kinetic wave theory approach for inhomogeneous media [5.37,38]. This model is an example of the third case given in Sect. 5.1 in which the different descriptions of the subsystems must be matched.

Let there be in the atmosphere a two-dimensional quasi-single-mode disturbance with arbitrary scales λ_z, λ_x. The hydrodynamic description is valid as long as the upward free wavelength is less than λ_z, or $l_0 + l_1 H \ll \lambda_z$, and for horizontal scales $l_0 \ll \lambda_x$. The horizontal gradient by the definition of a vertically stratified medium is zero. We introduce the height h that is the maximum height for the conditions considered in this problem to be valid. Therefore the atmosphere is split into two layers. In the upper layer $z \geq h$ the state of the disturbed gas is described by the Boltzmann equation. In the lower layer $0 \leq z \leq h$, the local equilibrium distribution function holds. In the upper layer we assume to be in the collisionless regime [5.38]. The lower regime is in the nondissipative hydrodynamics. For the matching condition we choose continuous distribution of particle fluxes. This simplest version of the model allows one to follow the derivations of the boundary conditions for the moments of the distribution function. This model has been successful in describing the stationary problem of vapor condensation [5.33,34]. Collisions are taken into account in three ways: (a) in a relaxation time approximation [5.39]; (b) by an expansion according to the number of multiple collisions [5.38]; (c) by the introduction of a narrow boundary layer with $K_n \sim 1$. The dissipative relaxation in the lower layer is taken into account as the viscosity and the thermoconductivity in a perturbation theory approach to the Navier–Stokes equations as in Sects. 2.5.

Let us consider the problem for the upper layer. The collisionless Boltzmann equation for an atmospheric gas in a gravitational field has the form is

$$f_t + v f_x + w f_z - f_w = 0 \tag{5.44}$$

where the designations are as before. The variables are dimensionless: the coordinates are measured in units of atmospheric scale H and the time in units of

$\sqrt{H/g}$. The general solution of (5.44) is given by the method of characteristics [5.38]. The characteristic equations have the form

$$t = \tau, \quad v' = v, \quad w' = w + t$$
$$\xi = x - vt, \quad \eta = z + t + w + \frac{w^2}{2},$$

that is, the arbitrary function of ξ, η, v', w' is the solution of (5.44)

$$f = \varphi\left(t', x - v(t - t'), h, v, w + (t - t')\right), \tag{5.45}$$

where $t' = w + t - w\sqrt{1 + 2(z - h)/w^2}$ and f is chosen so that at $z = h$ it coincides with the distribution function φ that defines the solution of the lower layer

$$\lim_{z \to h} f(t, x, z, v, w) = \varphi(t, x, h, v, w). \tag{5.46}$$

The condition of the continuity of the distribution function follows from its physical definition.

In a first order of the approximation in the amplitude σ the hydrodynamical distribution function is a linear combination of dynamical variables [5.32]. We use the variables δ, V, W, T, which were introduced in (2.1.) and which satisfy (2.19–21) except now the components of mass velocity are denoted by capital letters. We list here the linear version of the equations in these variables:

$$\begin{aligned} &\delta_t + V_x - W + W_z = 0 \\ &V_t + \delta_x + T_x = 0, \\ &W_t + \delta_z + T_z - T = 0, \\ &T_t + (\gamma - 1)V_x + (\gamma - 1)W_z = 0. \end{aligned} \tag{5.47}$$

One can see immediately, by substitution into the Boltzmann equation, that the distribution function to first order in amplitude leads to (5.47) and is equal to

$$\varphi(t, x, z, v, w) = f_0\left[1 + \delta + Vv + Ww + T\left(\frac{v^2 + w^2}{2} - 1\right)\right]. \tag{5.48}$$

Here f_0 is the equilibrium Maxwell–Boltzmann distribution (Sect. 4.1). For the mode representation of φ we need, as before, the explicit relations between the dynamical variables similar to (2.36). Let us write them for a single-mode long wave

$$\begin{aligned} W &= -\frac{1}{c_n}\Theta_t^n(x, t)e^{z/2}\sin k_n z, \\ V &= -\int_0^t \Theta_x^n(x, \tau)d\tau c_n[D_1^n \sin k_n z + D_2^n \cos k_n z]e^{z/2}, \\ \delta &= \Theta^n(x, t)[B_1^n \sin k_n z + B_2^n \cos k_n z]e^{z/2}, \\ T &= \Theta^n(x, t)[A_1^n \sin k_n z + A_2^n \cos k_n z]e^{z/2}. \end{aligned} \tag{5.49}$$

All the coefficients may be expressed in terms $A_{1,2}^n$:

$$B_1^n = \frac{A_1^n}{\gamma - 1} - \frac{1}{c_n}, \quad B_2^n = \frac{A_2^n}{\gamma - 1}$$

$$D_1^n = \frac{\gamma A_1^n/(\gamma - 1) - 1/c_n}{c_n}, \quad D_2^n = \frac{\gamma A_2^n}{\gamma - 1} \tag{5.50}$$

if $\Theta^n = \Theta_0^n e^{i(kx - \omega t)} + $ c.c. then

$$A_1^n = \frac{i(\gamma - 1)}{\omega} \frac{(2 - \gamma) + 4\gamma k_n^2 + 2(2 - \gamma)\omega^2}{(2 - \gamma)^2 + 4\gamma^2 k_n^2},$$

$$A_2^n = -i(\gamma - 1) \frac{4k_n}{\omega} \frac{1 + (\gamma\omega^2/\gamma - 1)}{(2 - \gamma)^2 + 4\gamma^2 k_n^2}.$$

The functions Θ^n satisfy the equation that is the corollary of (2.28) or the system (5.47)

$$\Theta_{tttt}^n - \gamma \Theta_{ttxx}^n + \frac{(\gamma - 1)\Theta_{tt}^n}{c_n^2} - (\gamma - 1)\Theta_{xx}^n = 0.$$

For the derivation of the boundary condition for the basis functions $Z = \sin(k_n z)e^{z/2}$, we substitute (5.49) into (5.48) taking into account (5.50), and then also into the matching condition (5.46). Using the explicit form of the distribution function (5.45), the transmission function of interaction with the upper layer, we calculate the derivative of w with respect to z. Multiplying then f by w, integrating over v, w, and taking the limit $z \to h$, we shall find the vertical hydrodynamic (mass) velocity

$$W_z \bigg|_{z=h} = \lim_{z \to h} \frac{\partial}{\partial z} \frac{1}{\varrho_0} \int_{-\infty}^{\infty} wf\, dv\, dw.$$

The results contain a term that is proportional to $Z(h)$, and since W_z is proportional to Z_z, after direct but awkward calculation, we get the simple result

$$Z_z(h) = \frac{Z(h)}{\gamma}. \tag{5.51}$$

The condition (5.51) represents the boundary condition of the third kind for mode functions that determine the discrete vertical wavenumber $k_z \equiv k_i$ spectrum for internal thermospheric waves. We emphasize that the coefficients in (5.51) do not depend on the mode number n.

Three values of k_n and corresponding propagation velocities are listed in Table 5.1. For comparison, the values calculated for other possible conditions used in [5.40-42] are given as well. It is interesting to note that the largest first mode velocity found from (5.51) greatly exceeds the velocity found when the boundary condition is $T_z = 0$ that is also compatible with (5.46).

Table 5.1. Single – mode internal wave parameters

$T_z\vert_{z=h} = 0$		$\gamma W_z\vert_{z=h} = W\vert_{z=h}$		Nonlinear constants	
k_n	c_n^2	k_n	c_n^2	$\Phi_{0n}\cdot 10^{-2}$	$\Phi_n\cdot 10^{-2}$
0.31	0.96	0.10	1.28	7.1	6.9
0.53	0.63	0.45	0.74	6.5	6.2
0.82	0.36	0.77	0.40	4.8	5.5

The formula for $Z_n(z)$, the condition (5.51), the expression for W, V, δ, T in (5.49,50) at $z \leq h$ together with the corresponding moments of the distribution function at $z > 0$ determine the single-mode internal wave in the assumed version of the model. The model reproduces the oscillating behavior of the wave field in the interval $z \in [0, h]$ and gives the exponential damping $\sim e^{-z}$ in $z \gg h$. We remind the reader that the hydrodynamical model of an internal wave in a boundless atmosphere yields the exponential growth $\sim e^{z/2}$. The crucial physical cause for amplitude decrease in rarefied atmosphere layers is the gravitational field action. The gravitational field returns the perturbed particles to before the moment of momentum transfer to other particles. The restriction on the choice of the matching height is not too limiting since a doubling in the wavelength requires only a ten percent shift in h. One can also derive the equations with a matching height dependence on k. In the linear theory the superposition principle is valid; in a weakly nonlinear theory superposition in this k dependence is allowed as approximation. The drawback of a model with a collisionless regime at $z > h$ is the neglect and of relaxation dissipative effects. The first step in the derivation is the approximation of the relaxation time in the upper layer [5.32]. Equation (5.44) is modified,

$$f_t + vf_x + wf_z - f_w = -\frac{f - f_0}{\tau_p},$$

where τ_p is the mean time of transition to the equilibrium distribution f_0 [5.31]. The modulus of the collision frequency $\nu_p = 1/\tau_p$ is equal to the minimum of the eigenvalues of collisionless regime at $z > h$ is the neglectimg of relaxation the linearized collision operator [5.32]. Such a model collision integral works well in the construction of interpolation formulas [5.31]. Going to $\xi = (f - f_0)/f_0$ we get the equation for ξ:

$$\xi_t + v\xi_x + w\xi_z - \xi_w = -\frac{\xi}{\tau_p}.$$

Using the same characteristics as in (5.45) and the condition (5.46) we go to

$$f = \exp[-(t - t')/\tau_p]\varphi\left(t', x - v(t - t'), h, v, w + (t - t')\right). \tag{5.52}$$

Let us consider further for simplicity the long waves approximation. The equation for the vertical velocity projection W has the form (5.3)

$$W_{ttzz} - W_{ttz} + \frac{\gamma-1}{\gamma} W_{xx} = 0 .$$

Projecting the equation and transforming the results to the form of (5.9) we find using (5.51):

$$e^{-h} Z_n(h)(W_z^+ - \gamma^{-1} W^+)_{tt} \Big|_{z \to h+0}$$
$$- \frac{(Z_n, W)_{tt}}{c_n^2} - \frac{(\gamma-1)(W, Z_n)_{xx}}{\gamma} = 0 . \tag{5.53}$$

Equation (5.53) after the substitution of a single-mode wave (5.49) and the boundary value W^+ found by (5.52) and after differentiating with the limiting transition, gives the equations for Θ^n:

$$\Theta_{tt}^n - c_n^2 \Theta_{xx}^n + \beta_n \Theta_t^n = 0 . \tag{5.54}$$

Equation (5.54) is referred to as the telegraph equation in which $\beta_n \sim 1/\tau_p$ is the decrement of attenuation. An analogous equation describes sound damping in a medium with resistance proportional to velocity. Its solution decays exponentially with the time constant β_n^{-1}. The calculation of the constant β_n shows that the mode lifetime decreases with the mode number n. Accounting for the viscosity and thermoconductivity leads to the same behavior (Table 2.1).

The next order pertubation theory equation for the amplitude σ leads to two effects. Firstly the usual hydrodynamical contribution that is generated by nonlinear terms in the system (5.47) appears. Secondly, the modification of the local equilibrium distribution function (5.48) through the boundary conditions influences the form of f. In the atmosphere this effect is important because of the exponential growth in the lowest layer. Providing for nonlinearity, dispersion, dissipation, and relaxation we obtain the KdV equation with damping

$$\Theta_t + c\Theta_x + \Phi\Theta\Theta_x + M\Theta_{xxx} + d\Theta = 0 ,$$

the coefficients of which differ from those of Table 2.1. The nonlinear constants are calculated by explicit formulas which yield an additional change in the soliton velocity of up to 20% (Table 5.1). The notations are $\Phi_n = \Phi_{0n} + \Phi_{1n}$, where Φ_{0n} is a nonlinear constant for the mode n in a single-layer model without the influence of the kinetic regime $\sim \sigma^2$. The dispersion constant is not changed. The constant d contains the additional relaxation term $d = d_{n0} + \beta_n$ (5.53) and (2.72).

The direct multiple collision expansion of f has been studied [5.38]. There the perturbation of a neutral gas by motion of ions in a rarefied gas is investigated. The method of calculations is taken from [5.44] and allows one to assume a small number of collisions when calculating the collision cross-section. The application of such a technique is unwieldy, but allows one to find the terms that correct the approximation for the relaxation time.

The theory presented in this section allows generalization to the few mode case. Even at a fixed matching height, the condition of validity that is related to the variation in the horizontal scale, does not significantly change. The vertical profiles of disturbances generated by different modes in the Knudsen regime are similar and their scales vary little. For further development of the theory the most important step might be the introduction of a transition region described by a linearized Boltzmann equation [5.35,40,45,46]. The entire system of equations can be applied to other physical problems besides plasma physics and turbulence theory, for example, problems where the kinetic description is used in the transition to a collisionless regime.

6. Mean Field Generation by Waves in a Dissipative Medium

This chapter begins with the description of the dissipation and diffusion effects in the presence of waves (Sect. 6.1). We relate dissipation to the mean field which is generated by the medium perturbation in the range of zero wavenumber that appears due to resonant nonlinear interaction of waves with finite amplitudes (Sects. 6.1,2). Diffusion effects are examined by using the example of ambipolar diffusion of F-layer ionosphere plasma in the presence of an internal gravity wave in the upper atmosphere (Sects. 6.2,3). The generation of fine structures is considered with the influence of dissipation in Sects. 6.4,5.

6.1 Description of a Wave-Medium System by the Scales of Motion

A propagating wave in a real nonlinear medium generally disperses, dissipates, and is accompanied by the changes in the medium that are no less interesting theoretically than the waves themselves. Such phenomena are related to, for example, plasma heating by electromagnetic radiation and ionosphere effects of internal waves. In speaking of wave effects we often imply only their action on a medium and ignore a back reaction. However, the description of wave evolution in inhomogeneous media is related to the same questions. It is known that in wave propagation in these media a variety of instabilities may appear, for example, internal wave instability in a shear flow, strong damping and increase of electromagnetic waves in active regions, drift instability of plasma and so on. In these cases the medium acts as the initiator of the variations in the wave intensity. The concepts that can unify these perspectives of wave behaviors in a medium should be based on interaction theory – a dynamical picture with equal importance of wave and medium. Before the description is split, leaving only the dynamical variables of the interaction partners, one should begin with a description of the closed system. As was already said, the most important roles in energy exchange are played by relaxation and dissipation. Nonconservative behavior in the subsystem can become nonconservative behavior of the whole system.

The problem of the separate description of wave–medium interaction acquires a great significance since contributions of the motion of one and the same object are under discussion. Traditionally there are several levels of definition of a medium. The medium is considered as (1) an arbitrary stationary state – the invariant solution of the fundamental equations system; (2) an active element that induces the wave energy change; (3) a passive object that is changed by a propagating wave; (4) the medium taking part in interactions. The latter generalizing description has been discussed in the first part of this book. Namely, the dispersion branches with the fundamentally different wave scales or periods of motion can be considered in those aspects of interest to us. Thus a Rossby wave in relation to an internal wave, or a large scale internal wave in relation to a sound wave may be represented from an experimental point of view as the dynamical state of the medium. Indeed, the experimental setup designed for detection of short waves cannot see a long wave. The theoretical description taking the interaction into account should give the results in which the influence of short waves should be noticeable after a sufficient time interval, however.

Another description of the medium that usually does not enter into the scheme of wave dispersion branches is the limiting stationary state of the fundamental equation system. (The state can be associated with the classification scheme with $\omega = 0$.) Examples of such states are the circulation of the atmosphere and ocean and zonal flows [6.1,2]. In this case properties of the mathematical apparatus appear that are related to the choice of the basis functions which allow one to find the interaction of those states with the waves. Besides the direct dynamical relationship in the basis states, the wave interaction can lead to the appearance of new structural elements. The fine structures, whose appearance is intimately tied to the waveguide propagation are related to these elements. In this case a standing wave structure is superimposed onto a background state producing coordinate dependences on scales that were absent in the initial state. This brilliantly expressed property is displayed by internal waves due to their inducement of an orbital movement of fluid particles. Moreover, internal wave propagation can induce transfer phenomena orthogonal to the propagation direction [6.3–6].

To illustrate, we consider the problem of propagation of a multi-mode internal wave that interacts with a shear flow [6.7,8]. Internal waves in a stratified flow have been studied in the classic works by Kelvin, Helmholtz, Taylor and Goldstein. It was shown that the coincidence of the phase velocity of a wave with the fluid velocity gives rise to instabilities [6.3,9–11]. The influence of nonlinearity on the propagation of internal waves has been investigated in a series of works [6.12–16]. The evolution equation for wave trains was derived and its solutions were studied as flow parameter functions. The fundamental equation was shown to be the nonlinear Schrödinger equation. Note that the statements made on the stability based on analytic and numerical analyses are contradictory [6.16].

The effects we discuss here are caused by the interaction between internal wave modes and mean flow. For the flow we introduce a dynamical description and a definite vertical structure. The problem is in the derivation of an equation with a mutually slow evolution in the amplitudes of the introduced dynamical variables. As is clear from the previous discussions, the spectral components of the internal waves that give rise to a disturbance of large horizontal scale and therefore influence the flow, should be taken into account. Namely, this contribution is related to the so-called mean field whose dynamical functions are included naturally in the background (ground) state [6.3].

The basis of the theory, as in Sect. 5.2 are the incompressible fluid equations in the Boussinesq approximation (5.16). For specific nonlinear effects in the initial state we assume the simplest possible stratification structure in the density and flow velocity. We assume that the wave propagates in a channel of depth H and that the water is stratified exponentially. Excluding the pressure from the equations (5.16) we write the system in dimensionless form by the scheme of Sect. 5.2. Letting $u \to \tilde{u} = u_0 + u$ we get

$$\tilde{u}_{tz} + T_x - w_{tx} = (\tilde{u}w_x + ww_z)_x - (\tilde{u}\tilde{u}_x + w\tilde{u}_z)_z, \tag{6.1}$$

$$T_t + w\bar{T}_z = -\tilde{u}T_x - wT_z, \quad \tilde{u} = -\int_{-\infty}^{x} w_z dx. \tag{6.2}$$

The flux as a dynamic object is defined by the projection of the horizontal velocity that is equal to $u_0 = U(\beta t, \beta x)\Phi(z)$. The remaining dynamical variables of the flow W_0, T_0 are small, of order β^2. In zero order the system (6.1,2) is reduced to an identity after substitution of the flow functions. As the initial condition we choose this solution with zero W_0, T_0. Then, due to the linear relationships $W_0 = -T_{0t} = -\int U_x \Phi dz$ and within the time interval that does not exceed β^{-1}, the influence of these components is neglible. However, they can become important (they are assessed in Sect. 6.4, for other purposes). The interesting point in this assumed description is that the vertical structure of the flow is different from the wave mode field structure. If for the long wave range of internal waves the known basis $Z^\nu(z)$ is defined by (5.22) and the zero boundary conditions at $z = 0, H$, then for a dimensionless z and a constant Väsäla frequency N we have

$$Z^\nu(z) = \sqrt{2}\sin \pi \nu z,$$

where $\nu = 1, 2, \ldots$. The flow dynamical varaible $\Phi_z(z)$ is determined independently and will not be orthogonal to the basis vectors. The assumption is in the introduction of the nonorthogonal basis relying on the known closed system Z^ν. The new basis we will numerate in the sequence where the vector $\Phi_z = e$, labelling it by the index "0". The next is Z^1; the vector Z^2 will be excluded since $(\Phi_z, Z^2) \ne 0; 0; Z^3$; and so on. Therefore $\nu = 0, 1, 3, 4, \ldots$.

The derivation depends on a projection procedure with the chosen basis. An arbitrary state vector φ of the fluid perturbation described by (6.1,2) will be represented as

$$\varphi = \varphi_0 e + \varphi_\nu Z^\nu, \quad \nu \ne 2. \tag{6.3}$$

Because the system of Z^ν is closed the coefficients are

$$\varphi_0 = \frac{(\varphi, e) - \sum_\nu (\varphi, Z^\nu)(Z^\nu, e)}{1 - \sum_\nu (Z^\nu, e)^2}, \tag{6.4}$$

$$\varphi_\nu = (\varphi, Z^\nu) - \varphi_e(e, Z^\nu). \tag{6.5}$$

Description of a Wave-Medium System by the Scales of Motion   117

The scalar product is determined in this section as $(\varphi, \Psi) = \int_{-1/2}^{1/2} \varphi(z)\Psi(z)dz$. The coordinate system origin is placed in the middle of the channel depth. The indices "0" and "ν" will denote projection onto the basis vectors. The projection will require linear independence instead of orthogonality. Note that the generalization of the mode representation is important in nonselfconjugate problems which may be obtained after separation of variables. Substituting $\tilde{u} = U\Phi + u$ into (6.1,3) we introduce the variables of the internal wave u, w, T.

Let us consider possible interactions of wave trains and a stratified flow in propagation of internal waves in a waveguide. For the vertical velocity projection we shall take the expansion in which every harmonic is a sum over the basis Z^ν:

$$w = \sum_{n\nu} \left(A^{n\nu} e^{i(k_{n\nu}x - \omega_{n\nu}t)} + \text{c.c.} \right) Z^\nu(z) . \tag{6.6}$$

In this representation the possible resonant generation of higher harmonics that naturally appear in the range of long waves due to the linearity of the dispersion function is taken into account. $A^{n\nu}$ are functions of the slow variables βx, βt. The index $n = 0$ separates out the mean field with $k_{0\nu} = \omega_{0\nu} = 0$. Due to (6.2) approximately

$$u = -\sum_{n\nu} \left[\int_{-\infty}^{x} A^{0\nu} dx + \sum_{n=1}^{\infty} A^{n\nu} e^{i(k_{n\nu}x - \omega_{n\nu}t)} / ik_{n\nu} + \text{c.c.} \right] Z^\nu_z , \tag{6.7}$$

$$T = \sum_{n\nu} \left(\tau^{n\nu} e^{i(k_{n\nu}x - \omega_{n\nu}t)} + \text{c.c.} \right) Z^\nu(z) . \tag{6.8}$$

In (6.7) integrals are taken to within an accuracy of $O(\beta)$. Substituting (6.6–8) into (6.1,2) and projecting over the basis $e = \Phi_z$, Z^ν we get

$$U_t = \sum_\nu (UA_x^{0\nu})_x (\Phi Z_\mu)_0 + \sum_{n\nu} 2(A^{n\nu})_x^2 (Z^\nu)_{z0}^2 - \sum_{\nu\mu} (A^{0\nu} A_x^{0\mu})_x (Z_z^\nu Z^\mu)_0 ,$$

$$\frac{A_t^{0\nu}}{c_\nu^2} + \tau_x^{0\nu} + ik_{n\nu}\tau^{0\nu} = A_{tx}^{0\nu} + \sum_\mu (UA_x^{0\mu})_x (\Phi Z^\mu)_\nu$$

$$+ \sum_{\alpha\mu} \int A^{0\alpha} dx A^{0\mu} \left(\frac{1}{c_\mu^2} - \frac{1}{c_\alpha^2} \right) (Z_z^\alpha Z^\mu)_\nu + \sum_\mu UA^{0\mu}(Z^\mu \Phi)_{zz\nu} , \tag{6.9}$$

$$\frac{A_t^{n\nu}}{ik_{n\nu}c_\nu^2} - \frac{\omega_{n\nu} A^{n\nu}}{k_{n\nu}c_\nu^2} + \tau_x^{n\nu} + ik_{n\nu}\tau^{n\nu} = \omega_{n\nu}k_{n\nu}A^{n\nu}$$

$$+ ik_{n\nu}A_t^{n\nu} - i\omega_{n\nu}A_x^{n\nu} + \sum_\mu ik_{n\mu}(A^{n\mu}U)_x(\Phi Z_\mu)_\nu$$

$$+ \sum_{\alpha\mu} \left(\int A^{0\alpha} dx A^{n\mu} + \frac{A^{0\mu}A^{n\alpha}}{ik_{n\alpha}} \right) \left(\frac{1}{c_\mu^2} - \frac{1}{c_\alpha^2} \right) (Z_z^\alpha Z^\mu)_\nu$$

$$+ \sum_\mu UA^{n\mu}(\Phi Z^\mu)_{z\nu} ,$$

$$\tau_t^{0\nu} + A^{0\nu} = -\sum_{\mu} U\tau^{0\mu}(\Phi Z^{\mu})_{\nu} + \sum_{\alpha\mu}\left(\int A^{0\alpha}\, dx\, \tau_x^{0\mu} - A^{0\mu}\tau^{0\alpha}\right)(Z_z^{\alpha}Z^{\mu})_{\nu}$$
$$+ \sum_{\omega_{n\alpha}+\omega_{n\mu}=0}\left(A^{n\alpha}\tau^{n\mu} + A^{-n\alpha}\tau^{n\mu}\right)(Z^{\alpha}Z^{\mu})_{z\nu}, \qquad (6.10)$$

$$\tau_t^{n\nu} - i\omega_{n\nu}\tau^{n\nu} + A^{n\nu} = \sum_{\mu} ik_{n\mu}U\tau^{n\mu}(Z^{\mu}\Phi)_{\nu}$$
$$+ \sum_{\alpha\mu}\left(\int A^{0\alpha}\, dx\, \tau^{n\mu}ik_{n\mu} - A^{n\mu}\tau^{0\alpha}\right)(Z_z^{\alpha}Z^{\mu})_{\nu}.$$

In this system the terms that are proportional to β^2 are neglected. These equations take into account the interaction between the stratified flow and internal waves that are in resonance with disturbances in a range of $k \sim 0$, the mean field generation and the self-action of the modes. Here the two-dimensional multimode system for a single layer thermoclyne with constant Väsälä frequency (5.30) is contained as a special case. We note that the indices on the z-functions denote the projections by (6.4,5).

Let us separate out the interaction which is of interest in this section. Let only that internal wave mode be excited initially which has $n = \nu = 1$. Denoting $A^{11} = A$, $\tau^n = \tau$, $c_1 = c$ we write the system, neglecting the possibility of generation of any other modes,

$$U_t = 2|A|_x^2(Z^1)_{z0}^2,$$
$$A_t - i\omega A + ikc^2\tau_x - k^2c^2\tau = ik^2c^2\omega A - kc^2(kA_t - \omega A_x) + UA(\Phi_z Z^1)_{z1},$$
$$\tau_t - i\omega\tau + A = ikU\tau(Z^1\Phi)_1.$$

Expressing A in terms of τ and taking into account the dispersion relation $\omega^2 = k^2c^2/(1 + k^2c^2)$ by assuming that the nonlinearity and the dispersion parameters σ, β are small, we get the simplest system that describes the internal wave packet and the flow interaction:

$$U_t = n_U|\tau|_x^2, \qquad (6.11)$$

$$\tau_t + c_g\tau_x + \frac{i\omega''(k)\tau_{xx}}{2} = in_{\tau}U\tau, \qquad (6.12)$$

where the nonlinear constants $n_U = 2\omega^2(Z^1)_{z0}^2$, $n_{\tau} = -k(Z^1\Phi)_1 + \omega^2(Z^1\Phi_z)_{z1}/k$ determine the interaction. The results of the nonlinear coupling constants calculations are listed in Appendix 4. Iterating the system in small parameters, i.e., substituting $\tau_t \approx -c_g\tau_x$ into (6.11) and integrating we find

$$U = U_0 - \frac{n_U|\tau|^2}{c_g}. \qquad (6.13)$$

Substituting the result (6.13) into (6.12) we go to the nonlinear Schrödinger equation (NS):

$$\tau_t + c_g \tau_x + \frac{i\tau_{xx}\omega''}{2} = \frac{in_U n_\tau |\tau|^2 \tau}{c_g} + iU_0 n_\tau \tau \ . \tag{6.14}$$

The soliton solution of (6.14) has the form

$$\tau = b \exp i \frac{\left[U_0 n_\tau - k_0 c_1 n_U n_\tau / c_0 - k_0 \sqrt{n_U n_\tau \omega''/2c_0}\right] t + \sqrt{n_U n_\tau \omega''/2c_0} k_0 x}{\cosh b \left[x - (c_g + c_2)t\right]} ,$$

where $c_2 = 2k_0$, $c_1 = (k_0^2 + b^2)/k_0$. The phase and group velocities are now the functions of the nonlinear constants and U_0. The flow is also changed by the quantity $U - U_0 = n_U b^2 / \cosh b \left[x - (c_g + c_2)t\right]$.

Internal wave propagation in a flow leads to generation of other modes that are related to the wave packet with the same ω, k. Their superposition characterizes the change in the vertical disturbance of the medium structure. The generation of the packet-accompanying modes to within an accuracy of $\beta\sigma$ is described by

$$\tau_t^\nu + c_g \tau_x^\nu = iU \sum_\mu n_\mu^\nu \tau^\mu \ ,$$

where the generation speed is determined by the right-hand side. The calculation results listed in Appendix 4 show that the values of the nonlinear constants are changed in sign and decrease with number as $\sim 1/|\mu - \nu|^2$; therefore, the largest contribution is made by modes having the nearest values of μ to ν (if there is no mode in resonance with the flow, Sect. 6.4 and Ref. [6.3]). The effects of the influence of an internal wave on the flow are not completely contained in (6.13). The mean field generation terms $A^{0\nu}$ are superimposed onto these effects; these terms are studied in Sects. 6.4,5. We note that accounting for all the dynamical variables (including W_0, T_0) generates an additional term $\int_0^t U_{xx} \int_0^z \Phi dz$ in (6.11). The system (6.11,12) is transformed to the Zakharov system [6.17] which, as is clear from the derivation described here in detail, is the universal system for describing the dynamics of long and short wave interaction (Sect. 4.2). The relative value of the additional term is given by the ratio Φ/Φ'' and in the relationship between the scales x and t, which are determined by the parameters β and σ. Equation (6.12) is therefore valid if $(\int_0^z \Phi dz)_0 \sigma^2/\beta^2 \ll 1$.

6.2 Thermoactivity and Diffusion in the Presence of Waves

The equations that determine the mass and heat transfer in a medium of wave propagation is contained in a dynamical system. Molecular motion that is induced by a wave leads to a density and temperature redistribution that changes the gradients of these quantities and therefore influences thermoconductivity and diffusion. The determination of these motions according to their scales and specific times becomes yet more complex. These processes, usually periodic for waves, are now modulated by the changes that are characteristic of irreversible relaxation phenomena [6.18]. The presence of stochastic turbulent pulsations increase the contribution and complexity of irreversible processes [6.19,20].

The statistical ensemble of interacting waves, as is shown in Chap. 4, can model weak turbulence phenomena. The time behavior of the spectral densities and envelope amplitudes of a multi-wave system give a possible representation of the development of a turbulent state and the values for coefficients of effective heat and mass transport [6.21,22]. The mutual action of the large-scale (in relation to the stochastic motion) coherent wave and the transfer effects determine the wave-medium system dynamics.

In this section, in continuing to develop the theory of internal wave propagation in an incompressible fluid we discuss taking into account thermoconductivity and viscosity in the general equations (2.10–12) and in the statement of the problem. First of all we note that the mass continuity and thermoconductivity equations are taken care of by the equation of state (2.10) $\varrho = \varrho_0(1 - \nu\tau_T)$, which is valid to a very good degree of accuracy for water internal waves. These equations give, instead of the usual div $v = 0$,

$$\text{div } v = \nu\varphi\kappa\Delta T' . \tag{6.15}$$

The total temperature τ_T is now separated as: $\tau_T = T + T'$, $T' = T + \theta$ where T is the dynamical variable of the medium (mean field; Sect. 6.1) and θ is a wave variable for a relatively short wave. In the designation of Sect. 2.1 $\varphi(z) = (1 - \nu\bar{T})/(1 - \nu\bar{T} + \bar{p}\nu/c_p)$ is a quantity whose value for water is very close to unity. We shall set $\varphi = 1$. Excluding the pressure from the two-dimensional equations of motion we get, instead of (6.1),

$$u_{tz} - w_{tx} + \nu g T'_x - \eta(u_{zzz} - \frac{4}{3}w_{zzx})$$
$$= (uw_x + ww_z)_x - (uu_x + wu_z)_z . \tag{6.16}$$

In place of (6.2) we get (6.15) and the thermoconductivity equation yields

$$\frac{d\tau_T}{dt} = \text{div } \kappa \Delta T' \approx (\kappa T'_z)_z ,$$

where (2.13,14) are taken into account. In the viscosity and thermoconductivity terms we retain only the vertical derivatives: we assume that the vertical scales are much smaller than the horizontal ones.

The transfer coefficients η, κ in real media may vary by several orders of magnitude, depending on the level of turbulence [6.22,23]. Moreover, the level of turbulence itself, as follows from the analysis of the semiempirical equations, changes because of shear flow that can be caused by internal waves [6.3,24]. Following [6.24] we shall briefly characterize this effect with vertical contributions only. The initial dynamic system is expanded by the inclusion of the turbulent energy balance equation [6.22]

$$b_t + (v\nabla)b = \frac{\partial}{\partial z}K_{bz}\frac{\partial b}{\partial z} + \eta(w_x)^2 - \kappa g \tau_{Tz} - \frac{c_4 \eta b}{l_z^2} , \tag{6.17}$$

where $c_4 = 0.046$, l_z is the vertical turbulence scale. By the Kolmogorov closing hypothesis the turbulence exchange coefficients $\eta = l_z\sqrt{b}$, $\kappa = \kappa_\varrho l_z\sqrt{b}$, and the turbulence diffusion coefficient $K_{bz} = \kappa_b l_z\sqrt{b}$ are proportional to $\sqrt{b}$ where $\kappa_{\varrho,z} b$ are empirical constants. By the results of [6.24] for exponential stratification with a Väsäla frequency N, a long wave $u = u_0 \sin(k_n z)\cos(\omega t - kx)$ with the condition $\omega^2 \gg (A/2 - D)l_z/c_4$, $A = l_z k_n^2 u_0^2$, $D = \kappa_\varrho N^2 l_z$, and the solution of the problem is the sum of the slowly varying part b_0 and an oscillating b_1 such that $b_1 \ll \max b_0$, k_n is the vertical and k the horizontal component of the wave vector:

$$b_0(z,t) = l_z \left[A\cos^2(k_n z) - D \right] c_4^{-1} \tanh\frac{t}{\tau_0}, \tag{6.18}$$

$$b_1(z,t) = \frac{(Ac_4)^{1/2}}{l_z^{1/2}\omega} b_0(z,t) \cos 2(\omega t - kx). \tag{6.19}$$

When $N = 10^{-2}\,\text{s}^{-1}$, $l_z = 1\,\text{m}$, $k = 10^{-2}\,\text{m}^{-1}$, $k_n = 10^{-1}\,\text{m}^{-1}$, $u_0 = 10^{-2}\,\text{m/s}$, the typical time of increasing turbulence is found to be $\tau_0 = 1.5 \cdot 10^4$ s and the mean maximum turbulent energy density is equal to $10^{-4}\,\text{m}^2/\text{s}^2$.

In the opposite limit $\omega^2 \ll (A/2 - D)l_z/c_4$ the turbulent energy follows the time dependence of the internal waves:

$$b = \begin{cases} l_z[k_n^2 l_z u^2 - D]/c_4 & k_n^2 l_z u^2 > 0 \\ 0 & k_n^2 l_z u^2 < 0. \end{cases}$$

Propagating islands of turbulency appear. In correspondence with the Kolmogorov closing formulas the coefficients of transfer differ from $\sqrt{b}$ by a multiplicative factor, i.e., (6.18,19) give the space-time dependence of the turbulent viscosity and thermoconductivity.

These results are produced in the first order perturbation theory. Their derivation does not take into account the variation in the wave field structure due to energy-momentum loss in the evolution process. The authors of [6.24] derive the expressions for damping decrements as a function of k for long internal waves, in connection with turbulence in the upper ocean layer (also via perturbation theory) for the first three modes. The mathematics is restricted to the framework of a two-layer stratification model. The conclusion is that the damping is large in the sharp vertical alternation of the diffusion coefficients.

In [6.24,25] and in the scientific report of the Kaliningrad State Univerity (GR 75017466, inv.571162, 1977) and in *Orlova's* (Kaliningrad State University, 1983) and *Yurova's* (Kaliningrad State University, 1985) theses various mechanisms of internal wave action in the ocean thermoclyne were investigated. The assumptions (as in [6.24]) about a mechanism for generation of an oceanic mean field were based on the action of internal wave energy dissipation that is accompanied by the thermoclyne energy redistribution in the vertical direction. In the process of internal wave propagation, as was shown in previous chapters, wave modes with higher $k_z = k_n$ are generated and dissipate. This process leads to the appearance of the mean fields with large horizontal and small vertical scales. The details of such fine structure generation are described in Sects. 6.4,5.

The reaction of a wave to a gas or other impurity or admixture in the atmosphere or ocean, whose diffusion is described by (2.1), is similar to dissipation. The mass velocity v that is produced by a wave, as well as the gradients that enter the diffusion velocity V_a [6.21],

$$V_a = -\frac{n^2 m_b}{\varrho n_a} D_{ab} \left\{ d + k_{\mathrm{T}} \frac{\partial \ln T}{\partial r} \right\}, \qquad (6.20)$$

yield the dynamics of the impurity. In (6.20) $n = n_a + n_b$ is the total concentration of particles, D_{ab} is the diffusion coefficient, k_{T} the thermodiffusion ratio, T the absolute temperature and

$$d = \frac{n_a n_b}{\varrho \cdot n} \left[\frac{m_a F_b - m_b F_a}{\kappa_{\mathrm{B}} T} + (m_b - m_a) \frac{\partial \ln T}{\partial r} \right.$$
$$\left. + m_b \frac{\partial \ln(n_a)}{\partial r} - m_a \frac{\partial \ln(n_b)}{\partial r} \right] = -d_b \equiv d_a,$$

If $n_a \ll n_b$, $m_a \approx m_b = m$, $\varrho \approx m n_b$, and the temperature gradients are not large, then neglecting thermodiffusion we get

$$V_a = D_{ab} \left[\frac{F_b - F_a}{\kappa_{\mathrm{B}} T} + \frac{\partial \ln(n_a/n)}{\partial r} \right]. \qquad (6.21)$$

By means of (6.20,21) one can calculate the local variations of concentration and the integral impurity transfer across the waveguide if we substitute the wave variables v, n into the diffusion equation (2.1). We write it here for the concentration n_a, with some other details:

$$n_{at} + \mathrm{div}\, n_a v + \mathrm{div} \left\{ \frac{n_a D_0 (F_b - F_a)}{\kappa_{\mathrm{B}} T \cdot n} + \nabla n_a + n_a \nabla \ln n \right\}, \qquad (6.22)$$

where the coefficient $D_0 = n D_{ab}$ does not depend on the concentration.

The diffusion of a small admixture during wave transit will be investigated in the next section in an example of ambipolar diffusion in ionospheric plasma. In this case the Lorentz force acts on the electrons and ions. Obviously, all particles are in the gravitational field. The Lorentz force on the rarefied gas leads to magnetization of plasma ions: particles are twisted around magnetic field lines and collisions can cause a change only in the mean projection of the velocity along the field lines.

6.3 Reaction of Ionosphere Plasma to a Passing Internal Wave

An internal wave propagating in the thermospheric waveguide (Sect. 5.4) is usually a superposition of a few contributions of the form (5.49) at an altitude lower than $h \simeq 500$ km [6.26,27]. Above $z = h$ the wave function is represented approximately (5.45) [6.28]. Note that h is the matching height for the hydrodynamical and kinetic descriptions that is taken as the upper boundary of the thermospheric waveguide. The neutral gas dynamics induced by an internal wave is represented by the wind with horizontal velocity projection that reaches hundreds of meters per second, alternating in direction. Such a movement of the thermosphere, as is shown in the previous section, necessarily entrains the ion component along the magnetic field [6.29,30] We note the F-layer plasma transfer that is situated at "classical" thermospheric altitudes. The variations in the position of the maximum concentration are observed from ground ionospheric stations as well as space installations. These methods of investigating thermospheric movement are not direct, but are widely applied and are the only ones that give rather complete information about internal waves. One might say that the theory of thermospheric internal waves is not complete without the theory of ionospheric effects [6.31].

Ionospheric plasma evolution was studied in [6.32,33] mainly by means of numerical simulations of the diffusion equation solutions. The analytic representation of the solutions has been derived by perturbation theory. The small parameter used was the dimensionless internal wave amplitude. In [6.34] the validity of the single-ion approximation has been substantiated for the ionospheric F-layer and the elementary solutions of the electron diffusion equations are given. Here we propose a generalized approach for broad internal waves spectra with arbitrary primary wind [6.35-37].

The conditions of a quasineutral plasma and ion magnetization in a rarefied atmosphere at the altitude of an F-layer allow one to describe the plasma movement by means of a single one-dimensional equation [6.37]. Let n be the electron density, D the product of the ambipolar diffusion coefficient and $\sin^2 I$ where I is the magnetic inclination [6.33]. The recombination velocity $\beta(z) = \beta_0 \exp(-z/H)$ which depends on the height over the lower thermosphere boundary (80 km) is also introduced. The scale of the recombination coefficient height is chosen to be equal to the neutral gas scale for simplicity. A more exact $\beta(z)$ function is determined from the molecular nitrogen concentration and yields other, more correct, basis functions.

Let us consider a long internal wave with a horizontal scale of 200–2000 km. These are the waves that are generated in the thermosphere in magnetic substorms. They propagate globally and manifest themselves in variations in the ionosphere parameters with periods on the order of hours [6.33,38]. The horizontal projection of the gas mass velocity induced by such an internal wave is much higher than the vertical velocity and therefore makes the main contribution to ion drag along the magnetic field lines. Projecting the hydrodynamical velocity first onto the magnetic field and then the result onto the vertical direction, we get the expression for $U = u \sin I \cos I$. Furthermore, if we denote the power density of the ionization source by q_1 we can write, transforming (6.22),

$$n_t - \left[D_0 e^{z/H}\left(n_z + \frac{n}{2H}\right)\right]_z + (nU)_z + \beta_0 e^{-z/H} n = q_1 . \tag{6.23}$$

The ion height scale is twice the neutral gas scale [6.34].

The boundary conditions for the electron density are determined by the known decrease of n in the lowest region of the ionospheric layer and by the presence of the plasma flux from the plasmasphere above:

$$n(0, t) = 0 , \tag{6.24}$$

$$n_z(h_i, t) + \frac{n(h_i, t)}{2H} = A , \tag{6.25}$$

where h_i is the ionosphere boundary height that exceeds the thermosphere boundary height h. Equation (6.23) also allows one to construct the theory of the perturbed ionospheric F-layer, that is, to explain its formation under the action of the ionization sources and the external plasma flux, accounting for diffusion and recombination. We will choose as the initial conditions for the calculations of the ionospheric response to the mutual action of the thermospheric circulation and an internal wave the stationary solution of the problem (6.23–25). An ionization source and a thermospheric circulation wind give the regular undisturbed 24 hour variations in the ionospheric plasma concentration.

The solution of the problem includes a transformation of (6.23) to a form convenient for an expansion in a series of the amplitude. The substitution relies on the simplification of the boundary conditions and the exclusion of the time dependence from the coefficients by differentiating as well as on a transition to a dimensionless form. The transformation accounts for the main exponential part of the dynamics. The new coordinates and time are dimensionless too:

$$z' = -2\mu^{-1} H \exp(-\frac{z}{2H}) , \quad t' = \frac{t}{\tau} \tag{6.26}$$

$$n = n' \exp(-\frac{z}{2H}) + \mu \int_0^{z'} \frac{f \, dz'}{2D_0} , \tag{6.27}$$

$$U = f(z, t) \exp(\frac{z}{2H}) . \tag{6.28}$$

In these transformations $\mu = H^{1/2} D_0^{1/4} \beta_0^{1/4}$ is the space and $\tau = 2H/D_0^{1/2} \beta_0^{1/2}$ the diffusion scale.

The substitution of (6.26–28) into (6.23) gives

$$n'_{t'} - n'_{z'z'} + \frac{(z')^2 n'}{4} + Wn' = \tau q_1 (z')^{-1} \times \exp\left(-\mu \int_0^{z'} \frac{f \, dz'}{2D_0}\right) , \tag{6.29}$$

$$W = \frac{\tau f_{z'}}{\mu} + \mu \int_0^{z'} \frac{f_t \, dz'}{2D_0} + \frac{\tau f^2}{4D_0} . \tag{6.30}$$

The boundary conditions (6.26,27) are also transformed to new variables:

$$n'(-\frac{2H}{\mu}, t') = 0, \tag{6.31}$$

$$n'_{z'}\left[-2\mu^{-1} H \exp(-\frac{h_i}{2H}), t'\right] + \frac{\mu f n'}{2D_0}$$

$$= \mu A \exp\left(\frac{h_i}{H} - \mu \int_0^{z'(h_i)} \frac{f dz'}{2D_0}\right). \tag{6.32}$$

The vertical structure of an internal thermospheric wave is such that a function f inside the interval $[0, h]$ is closed to superposition of the functions $\sin k_n z$. During the increase of z from h up to $h_i \approx h + 2H$ the function f decreases approximately by $e^{5/2}$ as U decreases by $e^{-z/H}$. Therefore we neglect the second term in (6.32) and the integral in the exponent in the right-hand side, $z'(h_i) \approx 0$. The condition (6.32) is significantly simplified:

$$n'_{z'}(0, t') = A'. \tag{6.33}$$

In reality, the exponent in the argument is near zero and A' should be chosen from experimental data.

The functions of the wave field are now concentrated in the potential function W. Let us write f in dimensionless form and introduce an amplitude parameter σ: $f = \sigma\sqrt{gH} f'$. Then the terms in (6.30) contain the following parameters. The first $\tau\sigma\sqrt{gH}(4\mu)^{-1}\sin 2I$ is the ratio of the wave amplitude with multiple $\sin 2I$ that depends on the magnetic inclination to the characteristic diffusion rate μ/τ at the lower boundary of the thermosphere. The second one contains the amplitude σ and the dispersion parameter λ_z/λ_x, appearing due to the derivative of f with respect to t. Lastly, the third one is proportional to the square of the amplitude. More exactly, it is proportional to $\sim \sigma^2 gH\tau/4D_0$. Thus, at small amplitudes one can use perturbation theory. The evaluation follows directly from the explicit form of the dimensionless parameters. One finds that waves that induce gas velocities of one hundred m/s at the height of the F2 layer are small, i.e., that those are moderate in the geophysical classification scheme. We note that the wave action was partly accounted already in the transform (6.29) and therefore it appears in the $W = 0$ case. The perturbation theory approach with which we shall derive the formulas for the reaction of the ionosphere to internal waves is, in essence, the nonstationary Dirac perturbation theory which is used in quantum mechanics.

We consider the nonperturbed problem with $f = 0$. Thus we introduce a stationary state that may be used as the initial condition and the basis of the perturbation theory. First we study the eigenvalue problem with the equation

$$-n'_{z'z'} + \frac{(z')^2 n'}{4} = \lambda n'$$

and the boundary conditions

$$n'(-\frac{2H}{\mu}, t') = 0 ,\tag{6.34}$$

$$n'_{z'}(0, t) = 0 .\tag{6.35}$$

From the boundary problem (6.34,35) it is convenient to express the solutions of (6.23) in terms of the Whittaker functions $V(\lambda, z')$ [6.39]. The substitution of the eigen functions $\Phi_n(z') = N_n V(\lambda, z')$ in the boundary condition (6.35) determines the spectrum $\lambda_n = -2n - 1/2$, $n = 0, 1, 2, \ldots$. The condition (6.34) is fulfilled within "physical" accuracy: the increase in the argument leads to a very rapid decrease in the Whittaker function, as $\exp[-(z')^2/4]$, that simulates the lower boundary of the ionosphere layer. N_n is a normalizing constant.

The stationary state $V(0, z')$ does not belong to the spectrum and should satisfy the conditions (6.31,32). The complete solution is therefore represented by a sum of the stationary state $V(0, z')$, the homogeneous equation with the solution of nonzero potential W, and the solution of the inhomogeneous equation with zero W.

The first order perturbation theory expression in wave amplitude is the nonstationary equation (we omit the primes below)

$$n_t - n_{zz} + \frac{z^2 n}{4} = -WV(0, z) + q .$$

This equation we solve by the Fourier method with the basis $\Phi_k(z)$

$$n = \sum_{k=0}^{\infty} \Phi_k(z) e^{\lambda_k t} n_k(t)$$

$$q = \sum_{k=0}^{\infty} q_k(t) \Phi_k(z) \tag{6.36}$$

$$-WV(0, z) = -\sum_{k=0}^{\infty} W_k(t) \Phi_k .$$

For the coefficients we get

$$n_k = \int_0^t [q_k(t_1) - W_k(t_1)] e^{-\lambda_k t_1} dt_1 ,$$

where

$$W_k(t) = \int_{-\infty}^{0} V(0, z) W \Phi_k(z) dz ,\tag{6.37}$$

$$q_k(t) = \int_{-\infty}^{0} q(z, t) \Phi_k(z) dz .\tag{6.38}$$

Using the properties of the Whittaker functions and the boundedness of the functions f we expand the integration in the scalar products up to $-\infty$. Finally,

$$n = V(0, z) + \sum_k \Phi_k(z) e^{\lambda_k t} \int_0^t e^{-\lambda_k t_1} (W_k + q_k) dt_1 . \tag{6.39}$$

The calculations were done by means of (6.26–28) and (6.36–39) to illustrate and test the method. The function $q(z,t)$ was chosen in the form of the "Chapman layer" [6.30]

$$q = q_0 \exp\left[1 - \frac{z - z_0}{H} \exp(-\frac{z - z_0}{H})\right] \varphi(t)$$

where z_0 is the height of the maximum and q_0 is the rate of ion production at the maximum. For simplicity we put $\varphi = 0$ at night and $\varphi = 1$ during the day. It was also assumed that in the thermosphere there is a background wind with a 24–hour period $u_0 = u_{00} e^{z/H} \sin \Omega t$. The action of the ionization source and the wind determine the 24–hour electron density behavior and the maximum height h_m as well as the variation of the maximum concentration n_m without the wave disturbance. The variations for F2 layer are shown in Fig. 6.1 by the dash-dot and dashed lines, respectively. The solid lines in the figure show the behavior of h_m and n_m after the internal wave disturbance. The internal wave was taken as a superposition of four modes. The calculation parameters were chosen as follows: $D_0 = 2 \cdot 10^6 \, m^2/s$, $\beta_0 = 4 \cdot 10^{-3} s^{-1}$, $A = 1.5 \cdot 10^{12} m^{-2} s^{-2}$, $q_0 = 10^9 \, m^{-3}$, $u_{00} e^{h/H} = 100 \, m/s$, $z_0 = 100 \, km$.

The condition of validity for the application of perturbation theory can be obtained from the estimation of the first term in W determined by the wave amplitude. In [6.36] formulas are derived without perturbation theory (in σ) but it is assumed that the vertical gradient is small. By means of these formulas one can find the effects of the weakly oscillation in z, but as a rule large in amplitude, first mode [6.40] and in this way relax the restrictions on the amplitude. Numerical calculations of the matrix elements W_k under the chosen moderate conditions give the values that lead to the relatively small contribution to the variation in h_m, n_m. The values of the variations are typical of experimental values [6.32,41].

6.4 Interaction of Internal Waves with a Dissipative Medium

6.4.1 Temperature Field Instability Induced by an Internal Wave

In Sect. 6.3 it has been shown that the medium turbulence level may be significantly influenced by the wave that induces shear flow. The level of turbulence determines the transport coefficients. Under real geophysical conditions primary turbulence generated by the shear flows is always present [6.23]. The values of the effective turbulent viscosity and thermoconductivity coefficients may vary up to a few orders of magnitude. Thus, if the molecular thermoconductivity coefficient for water is equal to $\kappa_m \approx 1.5 \cdot 10^{-7} \, m^2/s$ [6.3], then the effective coefficient for turbulent medium is $\kappa = \kappa_t = \kappa_m/\delta_1$, where $\delta_1 = 10^{-2} - 10^{-3}$ [6.23,24]. In spite of the very large values of the effective transport coefficients

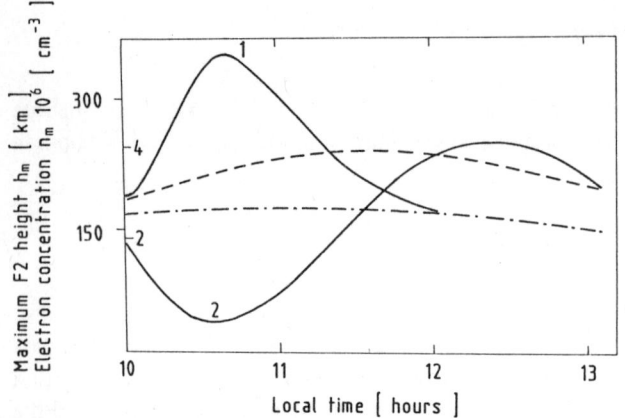

Fig. 6.1. Response of the ioniosphere to an internal gravity wave (*solid line*). Maximum electron concentration in the F2 layer (**1**) and maximum height (**2**) as functions of time. *Dashed* and *dash–dot* lines – unperturbed daytime variations

compared with molecular ones, the dimensionless dissipation parameter may be rather small for internal waves. If a wave is defined by a wavenumber k and a frequency ω then the parameter is $\varepsilon = \kappa k^2/\omega$. If the wavenumber k changes between 1 and 10^{-3} and $\omega \sim 10^{-1} - 10^{-3}$ s^{-1}, the parameter ε is between 10^{-1} and 10^{-7} even if δ_1 is at a maximum.

We can use the smallness of the parameter ε and try to state the tendency of a wave's influence on a thermoconductive medium by expanding the solution in a series in ε. To simplify the description we temporarily neglect the viscosity. The value of ε sets the degree of the energy exchange between the disturbances of the medium and its smallness allows one to use a two-scale expansion.

First we write the equations of interaction between an internal wave and the medium. The presence of the small parameter in the equations provides the rigorous basis for the separation of the dynamical variables in the wave-medium system. We consider (6.15–17) for the temperature perturbation and the velocity components u', w' in the two-dimensional problem. The dimensionless variables $t' = t\omega$, $x' = kx$, $z' = kz$, $v' = v\omega/k$ are introduced. We set to within an accuracy of first order in ε

$$T' = \Theta + \varepsilon T, \quad v = v^{\Theta} + \varepsilon v^T, \tag{6.40}$$

$$v^T = (u^T, w^T), \quad v^{\Theta} = (u, w). \tag{6.41}$$

Substituting (6.40,41) into (6.17) we get, omitting the primes,

$$(w + \varepsilon w^T)\bar{T}_z + \frac{d\Theta}{dt} + \frac{\varepsilon dT}{dt} = \varepsilon \Delta(\Theta + \varepsilon T).$$

Setting the functions in the same powers of ε equal to one another, we have in the zeroth and first orders

$$\frac{d\Theta}{dt} + w\bar{T}_z = 0,\tag{6.42}$$

$$T_t + uT_x + wT_z = \Delta\Theta - w^T\bar{T}_z.\tag{6.43}$$

From the continuity equation it follows that [see definition of ν in (2.10)]

$$\nu\left[T_t(w + \varepsilon w^T) + \frac{d\Theta}{dt} + \frac{\varepsilon dT}{dt}\right] = \operatorname{div} \boldsymbol{v}\Theta + \varepsilon \operatorname{div} \boldsymbol{v}^T.\tag{6.44}$$

Combining (6.42) and (6.44) we get in the same approximation

$$u_x + w_z = 0,\tag{6.45}$$

$$\operatorname{div} \boldsymbol{v}^T = \nu\Delta\Theta.\tag{6.46}$$

Representing the pressure in the form of (6.40) $p = \bar{p} + p' = \bar{p} + p^\Theta + \varepsilon p^T$, we write the corollaries of the equations of motion

$$\frac{\varrho_0 d\boldsymbol{v}^\Theta}{dt} = -\nabla p^\Theta + k^{-1}\varrho_0\nu\Theta\boldsymbol{g},\tag{6.47}$$

$$\boldsymbol{v}_t^T + (\boldsymbol{v}^\Theta, \nabla)\boldsymbol{v}^T + (\boldsymbol{v}^T\nabla)\boldsymbol{v}^\Theta = -\nabla p^T + k^{-1}\nu T\boldsymbol{g}.\tag{6.48}$$

Excluding the pressure from (6.47) we have

$$\frac{\partial}{\partial z}\frac{du}{dt} - \frac{\partial}{\partial x}\frac{dw}{dt} = -\frac{\nu g k}{\omega^2}\frac{\partial\Theta}{\partial x}.\tag{6.49}$$

Equations (6.42,45,49) comprise the classic closed system for the wave variables u, w, Θ of an internal wave in a two-dimensional incompressible temperature-stratified fluid (Sect. 5.2). The linear version of this system may be transformed to the dimensionless partial form of (5.17) for the vertical velocity components

$$(\Delta w)_{tt} + \frac{kg\nu\bar{T}_z}{\omega^2}w_{xx} = 0,$$

where $N^2 = g\nu\bar{T}_z k/\omega^2$ is the square of the Väsälä frequency. For the dynamical variables u^T, w^T, T of the medium disturbance generated by the thermoconductivity we get from (6.43,46) and from the exclusion of p^T from (6.48)

$$u_x^T + w_z^T = \nu\Delta\Theta,\tag{6.50}$$

$$w_{tx}^T - u_{tz}^T - \frac{g\nu k T_x}{\omega^2} = M,\tag{6.51}$$

$$T_t + uT_x + wT_z = \Delta\Theta - w^T\bar{T}_z,\tag{6.52}$$

where $M = [(\boldsymbol{v}^T, \nabla)u + (\boldsymbol{v}^\Theta, \nabla)u^T]_z - [(\boldsymbol{v}^T\nabla)w + (\boldsymbol{v}^\Theta, \nabla)w^T]_x$. Dropping u^T, we go to

$$\nabla w_t^T - \frac{g\nu k T_{xx}}{\omega^2} = \nu\Delta\Theta_{zt} + M_x.\tag{6.53}$$

The system (6.50–52) or (6.52,53) describes the internal wave action on the ground state of the medium to first order perturbations in ε. Solving these equations one can derive the increment of increase in the given spectral contributions due to thermoconductivity. By this method the singularities may be detected, and they may be compensated for within the framework of the nonsingular perturbation theory (Sects. 2.4,5). It is namely within this approach that the action of the medium on the long wave has been considered in Sects. 2.4,5 and which describes the wave damping.

One can eliminate w^T by applying the operator $\Delta \partial/\partial t$ to (6.52) after the multiplication by $\bar{T}_z^{-1}$:

$$\Delta \frac{\partial}{\partial t}\left(\frac{T_t + uT_x + wT_z}{\bar{T}_z}\right) = \Delta\left(\frac{\Delta \Theta_{zt}}{\bar{T}_z}\right) - \nu \Delta \Theta_{zt} - \frac{g\nu k T_{xx}}{\omega^2} - M_x \,. \quad (6.54)$$

Thus we have derived the equation for the contribution of T to the mean field of the temperature. This term appears due to the thermoconductivity because the source in (6.54) in dimensional form is proportional to the thermoconductivity coefficient. The expression of M as a function of T, Θ is obvious but awkward, and is determined from (6.50) and (6.52).

The statement of the problem includes the same boundary conditions as the total temperature perturbation T': $T|_{z=0,H} = 0$. For an incompressible fluid, in the linear approximation without dissipation this follows from the condition $w|_{0,H} = 0$ and the above equations. The initial condition in the problem of the generation of the mean field is zero: $T(x,z,0) = 0$.

The additional contribution to the baseline temperature T that appears from the interaction with an internal wave varies slowly with time but it may contain terms oscillating rapidly with coordinate z. The functions that describe the internal waves enter into the coefficients of (6.54) as well as into the right-hand side. Therefore the solution of this equation must contain oscillations at the fundamental wave frequencies as well as at combination frequencies. Through the combinations of $e^{\pm ikx}$ we get the contribution of the "clean background" that does not depend on x (6.9,10).

We now turn our attention to this last term. We shall calculate the rate of its growth in the approximation of weak nonlinearity for the given single-mode plane wave as the source. Due to the linear coupling, in analogy with the atmospheric case (5.49), we write

$w = \sigma \tilde{w}(z) e^{i(\omega t - kx)} + \text{c.c.}$

$u = \sigma \tilde{u}(z) e^{i(\omega t - kx)} + \text{c.c.}$

$\Theta = \sigma \tilde{\Theta}(z) e^{i(\omega t - kx)} + \text{c.c.}$ \quad (6.55)

where

$$\tilde{\Theta} = \frac{i\tilde{w}\bar{T}'(z)}{\omega}, \quad \tilde{u} = -\frac{i\tilde{w}}{k} \,. \quad (6.56)$$

We continue to develop the perturbation scheme. Expanding the variables in a series in the amplitude σ

$$T = T_0 + \sigma T_1 + \ldots \tag{6.57}$$

we get the equations for T_0, w_0^T, u_0^T:

$$-\Delta T_{0tt} - N^2 T_{0xx} + \Delta^2 \Theta_{zt} - \nu \bar{T}_z \Delta \Theta_{zt} = 0 , \tag{6.58}$$

$$w_0^T = \bar{T}_z^{-1}(\Delta\Theta - T_t) ,$$
$$u_{0x}^T = \nu\Delta\Theta - \bar{T}_z(\Delta\Theta - T_t) . \tag{6.59}$$

Investigation of (6.58) shows that the amplitude of T_0 at a given amplitude of Θ increases with the wave vector component ratio k_z/k_x. In a region with $k_z \gg k_x$, $k_z \gg 1$ one may neglect the term $N^2 T_{0xx}$. One should note that the unit of length is the reciprocal horizontal wave number $k_x^{-1} = k^{-1}$. The integration in (6.58) then becomes trivial

$$T_0 = \int_0^t \frac{\Theta_{zz} dt}{\sigma} = \frac{\tilde{\Theta}_n'' e^{i(\omega t - kx)}}{i\omega} + \text{c.c.} . \tag{6.60}$$

Now as follows from (6.59), $w_o^T = \nu\Theta_z \ll \nu T_0$, $u_0^T = \nu\Theta_x \ll \nu T_0$. An estimation of the nonlinear contributions of M in (6.54) in the next order perturbation theory gives a value that is much less than the nonlinear terms $uT_{0x} + wT_{0z}$. From the expansion in (6.57), taking into account all the approximations, we get the equation for T_1:

$$\sigma T_{1t} = -uT_{0x} - wT_{0z} . \tag{6.61}$$

Substituting the expressions for u, w from (6.55) and T_0 from (6.60) we shall come to the formula for T_1 that contains the increment of the additional temperature modulus. Neglecting the oscillating restricted terms we write the expression for the increasing contribution to the temperature field

$$T_1 = \frac{2\varepsilon\sigma^2 t(\tilde{\Theta}''\tilde{w})'}{\omega} .$$

Using the coupling of $\tilde{\Theta}$ and w via (6.56) and taking $\bar{T}'$ weakly depending on z outside the derivative, we get

$$T_1 = \frac{\varepsilon\sigma^2(\tilde{w}\tilde{w}'')' \bar{T}' t}{2} . \tag{6.62}$$

The rate of the irreversible medium temperature change is equal to the coefficient of t in (6.62). All the variables in (6.62) except the temperature are dimensionless. The vertical scale of the mean field disturbance is equal to half of the vertical scale of the mode under study. The total effect is a sum of the expressions in (6.62) if we neglect the mode interaction. The role of the terms with different n is not equal, as we shall now show.

The value of T_1 is proportional to κ and therefore is related to dissipation. The local density of the wave mode energy in the linear approximation is equal to the sum of the kinetic and potential energy and is expressed by [6.3].

$$E = K + \Pi = \frac{\varrho_0(u^2 + w^2)}{2} - \frac{g\varrho^2}{2\varrho_0'(z)} .$$

If $\tilde{w} = (a \sin k_n z)/2$ then when $\omega t - kx = 0$ the energy density on a unit surface area is

$$E_n = \varrho_0 a^2 h \left[\frac{k_n^2}{k^2} + \frac{N^2}{\omega^2} \right] . \tag{6.63}$$

Calculating the heating rate of the medium for the unit surface area and period

$$\delta E_n = c_p \int_{-H}^{0} \frac{T_{1n} dz}{t} = \frac{\varepsilon c_p k_n^2 a^2 \bar{T}'}{4} ,$$

and comparing it with (6.63), we can find the percentage of wave energy loss or the decrement of decrease. Of course the estimation formulas are given here for values of the variables restricted by all the assumptions. For instance, it is necessary that $\delta E_n \ll E_n$.

6.4.2 Nonsingular Perturbation Theory

We proceed to a deeper analysis with an asymptotic treatment, taking into account the noted singularities. Obviously, the wave may heat the medium and the degree of heating depends on the local stratification properties of the fluid ($T_{1n} \sim \bar{T}'$), the value of the thermoconductivity coefficient, the square of the amplitude and in particular, on the mode number. The presence in (6.62) of the third derivative with respect to z means that the rate of increase in T is proportional to the cube of n, since $k_n \sim n$. Therefore, the increase in the mode number leads to wave energy loss and mode excitation will be supressed.

The mode excitation may be directly due to the sources action or to the nonlinear generation mechanism that, according to widespread opinion, is more probable [6.3,4,42]. The generation of higher modes has been studied in Sects. 2.4,6 and 5.2,6.1. In this section it is described by the system (6.42,45,49) without taking into account dissipation effects. Now we shall try to unify the nonlinearity, dispersion and dissipation effects separating the interaction of the wave localized in a narrow spectral interval and the long wave perturbation with a small propagation velocity. The last will be identified as the mean field and the corresponding variables with the medium variables.

The task is to find the most important of all the parameter couplings. Then we shall try to guess the small parameter asymptotic representation. To this end we introduce the nonlinearity σ and dispersion β parameters. In a problem with dissipation we include the heat exchange intensity which we shall determine now, taking the viscosity into account: $\varepsilon = (\kappa + \eta)k^2/\omega$. The introduction of dissipation has been displayed partly in the statement that the processes of the nonlinear transformation are significantly influenced by the modes of large n that produce large gradients. In the first part of this section, in the discussion of the estimation formulas by the direct perturbation theory we found the characteristic value of the vertical scale $k_{n0}^{-1} \equiv \varepsilon_1$. It is this scale that determines the degree of energy exchange between modes through thermoconductivity and viscosity and allows one to estimate the magnitude of the vertical derivatives of the mean field. The use of the vertical derivatives scale follows from the presence of the resonant contribution in the nonlinear interaction [6.3,42,43]. The existence of the resonance is seen from the form of the operator in (6.54) or (6.58). Based on the dimensions and the empirical values of the parameters, we can choose the relationship between the parameters to be

$$\varepsilon_1 = \sqrt{\frac{\varepsilon}{\beta}} \,. \tag{6.64}$$

We emphasize that the specific real coupling is conditional to some degree and depends on the choice of the range of the wave train spectrum, the dispersion relation (the values of β and c_g) and the degree of primary turbulence (i.e., κ and η values). If the degree of turbulence is given by an amplitude (Sect. 6.2) then the coupling (6.64) may contain σ. We assume that the magnitude of ε_1 is on the order of σ, which means that the turbulence, amplitude and wave train forms are consistent.

The system of equations coincides with (6.15–17). We shall account for the increase in turbulence of the medium by the internal wave field by assuming that it changes together with the wave envelope or by (6.18), neglecting the small oscillations. Applying again sequential approximations, we shall neglect the turbulence dynamics, first regarding κ, η as constants and taking the dynamics into account only in the last step of the calculations.

To demonstrate the calculations explicitly we shall take the two-layer model with a constant Väsälä frequency in the upper layer. The lower layer we shall consider to be uniform ($N = 0$). Let the upper waveguide layer occupy the interval $[0, -h]$ and the lower be at $[-h, -H]$. The basis of the vertical coordinate functions is denoted as $Z_n(z)$, whose equation is written by separating the variables with $\varepsilon = \sigma = 0$. The matching conditions at the layer boundary and the zero conditions at $0, -H$ give the spectrum of the vertical wave number $k_z = k_n$

$$k \tan k_n H = -k_n \tanh k(H - h) \,. \tag{6.65}$$

Let us assume that the thermoclyne is excited by the internal wave that is localized in the first mode with $k_z = k_1$, $Z_1 = Z$ and that the mean field is given by a superposition of the remaining modes:

$$T = \bar{T}_z \sum_{n=2}^{\infty} T_n Z_n \,, \tag{6.66}$$

$$W = \sum_{n=2}^{\infty} W_n Z_n \,, \tag{6.67}$$

$$Z_n'' = k^2 \left(1 - \frac{N^2}{\omega^2}\right) Z_n \,. \tag{6.68}$$

The asymptotic representation (in the small parameter $\sigma \sim \beta \sim \varepsilon$) of the solution is chosen to be the wave and mean field sum

$$T' = \sigma A(\beta t, \beta x)\bar{T}_z Z e^{-i(kx-\omega t)} + \text{c.c.} + \sigma^2 T(\beta t, \beta x, \frac{z}{\varepsilon}),$$

$$W' = \sigma B(\beta t, \beta x) Z e^{-i(kx-\omega t)} + \text{c.c.} + \beta\sigma^2 W(\beta t, \beta x, \frac{z}{\varepsilon}). \tag{6.69}$$

To an accuracy of up to $\sigma^2\beta$ one may neglect the interaction between the generated mean field modes. We shall write the resonant or near resonant mode interaction equations, omitting the index n. We substitute (6.69) into (6.15–17), taking into account the mode contributions (6.66,67). After applying the transformation to eliminate the variables of vertical movement and projecting over the eigen subspaces of the Sturm–Liouville problem (6.65,68), we get

$$T_{tt} - c_n^2 T_{xx} + \nu_n T_t = D_1 |A|^2_{tt} + D_2 |A|^2_{xx}, \tag{6.70}$$

$$i[A_t + c_g A_x] + \frac{\omega''(k)}{2} A_{xx} = \mathcal{E}_1 \left(\mathcal{E}_2 \int_{-\infty}^{x} T dx - \int_{-\infty}^{x} T_t dx \right) A,$$

where $c_n^2 = N^2/(k_n^2 + k^2)$, $c_g^2 = N^2 k_1/(k_1^2 + k^2)^{3/2}$

$$D_1 = -c_n^2 \int_{-H}^{0} \frac{(Z^2)''' Z_n dz}{M_n N^2}$$

$$D_2 = 2c_n^2 \omega^2 \int_{-H}^{0} \frac{(Z^2)' Z_n dz}{N^2 M_n}$$

$$\mathcal{E}_1 = \omega^2 \int_{-H}^{0} \frac{Z^2 Z_n' dz}{2M_1 k c_n^2}$$

$$\mathcal{E}_2 = \frac{(\kappa+\eta)N^2}{c_n^2}, \quad \nu_n = (\kappa+\eta)(k_n^2 + k^2)$$

$$\omega = \frac{Nk}{\sqrt{k_1^2 + k^2}}, \quad M_n = \int_{-H}^{0} Z_n^2 dz.$$

When $\kappa + \eta = 0$ the system (6.70) becomes the Zakharov system (4.42). As we noted in Sect. 4.2, (4.42) has stable stationary solutions. That is why the wave can effectively interact with the resonant mode. In the first part of this section it was shown that the dissipative-resonant interaction is also effective. The inclusion of modes with similar n values may also be effective in the interaction. In Sect. 6.5 we shall give more detailed formulas. We add that the dependence of dissipative coefficients on a wave amplitude may magnify the nonlinearity of the system (6.70).

This section shows the possibility of comparing two types of perturbation theory. It is seen that the application of the nonsingular theory allows one to expand the validity of the approximate description over a wider time interval.

6.5 Mean Field Generation in a Dissipative Medium and Fine Structure of the Oceanic Thermoclyne

As was shown in Sects. 6.2,4 the quasiperiodic internal wave propagation to nonlinear interactions generates a disturbance whose spectrum is localized near the zero of the horizontal wave vector and frequency. The results of this interaction in plasma, fluid and gas physics [6.3,4,17,21,42–44] leads to the Zakharov system of equations and its modifications. The intrinsic inhomogeneity of the internal wave appears primarly in the induced field of the stratified shear flows that influence the medium turbulence (Sect. 6.2)[6.23,24] and secondly in the effective generation of modes with large vertical wavenumbers. The appearance of a mean field oscillating rapidly along the vertical direction and slowly along the horizontal axis in the ocean is interpreted as the fine structure of the ground state or the state averaged over the period of the wave field oscillations. It is the nonlinear mechanism that explains the broad fine structures in the ocean thermoclyne cited by *Miropolsky* and others [6.3,4,17,42,44,45] who criticized the linear superposition theory of *Gurret* and *Munk* [6.46]. The vertical scale of fine structures vary from centimeters (microstructures) to tens of meters. For example, in [6.4] a ~ 3 m scale is considered. The theory of fine structure formation is useful for illustration of the methodology of the solution of the propagation of nonlinear waves in a waveguide.

The authors of [6.3,4,42] noted the incomplete nature of the theory of nondissipative fine structure generation. They have discussed reversibility of the processes and the increases in perturbations at exact resonance to infinity. In this section we analyze the dynamics of the decay and formation of fine structure by the formulas for a dissipative turbulent medium. The obtained results, as noted above, are the further steps of the theory elaborated in [6.24,42,44].

One can distinguish two ways of a nonlinear wave reaction to a mean field: (1) direct generation through the initial wave and very long waves interaction (6.70); (2) generation of rapidly vertically oscillating (higher) modes due to second harmonics that in turn generate the mean field by (6.62). We shall first consider the first mechanism, splitting the system (6.70) without account of the back reaction to the wave [6.47].

Let the wave envelope in the first mode be nonzero in the region $t - x/c_g > 0$ and let it propagate without dispersion and damping. Then, in accordance with (6.70),

$$T_{tt} - c_n^2 T_{xx} + \nu_n T_t = f_{tt}\left(t - \frac{x}{c_g}\right),$$

$$f(\eta) = f_0 \Theta(\eta)\left[1 - \exp(\frac{\eta}{t_0})\right], \quad \eta = t - \frac{x}{c_g},$$

$$f_0 = A_0^2 \left(D_1 + \frac{D_2}{c_g}\right).$$

Here A_0 is the amplitude of the initial internal wave, Θ is the Heaviside function, t_0 is the time of the wave increase. The initial condition is naturally chosen to be zero for $T(0) = T_x(0) = 0$. We calculate the temperature at the point $x = 0$. We assume that at $t = 0$ the wave train has appeared inside the interval $x < 0$. Assuming that $\nu(t)$ represents the envelope of the wavetrain, the explicit form for T is

$$T = \frac{c_g}{2c_n} \exp\left(-\frac{1}{2}\int_0^t \nu_n d\tau_1\right) \int_0^t \exp\left(\frac{1}{2}\int_0^{\tau_2} \nu_n \tau_3\right)$$
$$\times \left\{ f'\left(\frac{(c_n + c_g)\tau_2 - x - c_n t}{c_g}\right) - f'\left(\frac{(c_g - c_n)\tau_2 - x + c_n t}{c_g}\right) \right\} d\tau_2 .$$

To estimate the magnitude of the effect we simplify the resulting formula (6.65) by setting $\nu_n = $ const. Integrating we have

$$T = \frac{f_0 c_g^2}{c_n} \left\{ \frac{1}{c_g \nu t_0 - 2(c_n + c_g)} \left[e^{-t/t_0} - \exp\left(-\frac{\nu c_g t}{2(c_g - c_n)}\right)\right] \right.$$
$$\left. - \frac{1}{c_g \nu t_0 - 2(c_g - c_n)} \left[e^{-t/t_0} - \exp\left(-t\left(\frac{c_n}{c_g t_0} + \frac{\nu}{2}\right)\right)\right] \right\} . \qquad (6.71)$$

We note immediately that the expression for T is still nonsingular at the resonance where $c_g \to c_n$. At the point $c_g = c_n$ we get

$$T = f_0 \left[\frac{1}{\nu t_0 - 4}\left(e^{-t/t_0} - e^{\nu t/4}\right) - \frac{e^{-t/t_0}}{\nu t_0}\left(1 - e^{-\nu/2}\right)\right] .$$

The mode amplitudes have been calculated by (6.65,66) and the summation over n gives the structure that qualitatively coincides with the figures of [6.3]. The parameters of the calculations were chosen such that $h = 100$ m, $H = 1000$ m, $\kappa = 1.5 \cdot 10^{-4}$ m²/s, $\eta = 10^{-3}$ m²/s.

Now we shall consider the second mechanism of a fine structure formation. Let the interaction of the basis wave with higher modes result in a perturbation in a mode n with an amplitude $\tilde{a}$. Using (6.63) we set the mode energy and the energy loss in a period equal, i.e., we shall assume that there is an approximate equilibrium between them that is reached at the amplitude $\tilde{a}$. Due to (6.63) we have

$$c_p \int_{-h}^0 \frac{T_{1n} dz}{t} = \frac{\varepsilon c_p k_n^2 \tilde{a}^2 \bar{T}'}{4k} .$$

If we put T_{1n} from (6.62) and the mode expression from (6.69) we get

$$(\Lambda - 1)k_n^2 = \frac{N^2}{\omega^2} , \qquad (6.72)$$

where $\Lambda = (c_p \kappa_m N^2 k^2)/(\omega^3 4\nu \varrho_0 h) = \kappa D(k_n^2 + k^2)/\omega\delta$, $D = c_p/\nu \varrho_0 h = 3 \cdot 10^4$, is determined by the physical constants of water and the thermoclyne depth.

The condition (6.72) defines the range of parameters in which energy transfer is effective. The generation of the n mode is not suppressed very much by dissipation although the dissipation is rather strong. Thus, because the coefficients (6.72) are positive, $\Lambda > 1$ and the value of $|T_t|$ must be large enough. The condition for Λ means that

$$\varepsilon N^2 > \omega^2 D^{-1} \ . \tag{6.73}$$

For the two-layer model of the primary vertical structure of the ocean that has been introduced in Sect. 6.4 in $N = 10^{-2}\,\mathrm{s}^{-1}$, $k_x = 2 \cdot 10^{-2}\,\mathrm{m}^{-1}$, $\delta = 10^{-3}$, $k \gg k_n$, the condition (6.73) gives $\omega \lesssim 2 \cdot 10^{-3}$. Therefore the effect exists in internal waves of periods in the range of hours and the rate of increase is determined by the value of the vertical wave number $k_n = N(\Lambda - 1)^{-1/2}/\omega$. If $\omega = 10^{-3}\,\mathrm{s}^{-1}$, $n = 30$, then the rate of increase of the mean field is equal to

$$\max T_1' = \frac{\kappa N k_n^3 \tilde{a}^2}{2\delta \nu g} = \frac{3 \cdot 10^{-7} \cdot 5 \cdot 10^{-4} \tilde{a}^2 \cdot 10^6}{2 \cdot 10^{-3} \cdot 1.5 \cdot 10^{-4} \cdot 10} = 50 \tilde{a}^2 \tag{6.74}$$

and corresponds to a vertical structure with a scale of 3 m. For a dimensionless amplitude $\sigma = w_{\max}/c_f = \tilde{a}$, such that $\tilde{a}^2 \sim 10^{-3}$ and rate of increase is equal to $5 \cdot 10^{-2}$ during the period. The addition of modes with similar values of n may increase the effect. Viscosity also increases T_1'. It may be shown that to include viscosity effects it is enough to change $\kappa \to \kappa + \eta$ in (6.74).

Thus, if in the thermoclyne the wavetrain with parameters that are in the range of the required conditions is propagated, the effects of resonant exchange can produce the fine structures. The mechanisms that were discussed in this section are complementary. If the first one is effective in the region of long wavelengths, then the second is good for short waves. Dissipation effects in the second case cancel the singularities and lead to remaining fields that decrease with a velocity which is much less than the rate of increase since the additional turbulence goes away together with the oscillating train and moving mean field.

Appendix 1

Atmospheric Waves over a Rotating Planet

The system of equations (2.25,26) may be written in the form

$$A\delta + BT = F, \tag{A1.1}$$
$$C\delta + DT = G$$

where the operators A, B, C, D are determined by the equalities

$$A = \frac{\partial^4}{\partial t^4} + 4\Omega^2 \frac{\partial^2}{\partial t^2} + 4(\boldsymbol{\Omega}, \boldsymbol{n})(\boldsymbol{\Omega}, \nabla) + 2(\boldsymbol{m}, \nabla)\frac{\partial}{\partial t} + \frac{\partial^2}{\partial t^2}(\boldsymbol{n}, \nabla),$$

$$B = -\frac{1}{\gamma - 1}\frac{\partial^4}{\partial t^4} - \frac{4\Omega^2}{\gamma - 1}\frac{\partial^2}{\partial t^2} + 4(\boldsymbol{\Omega}, \boldsymbol{n})(\boldsymbol{\Omega}, \nabla) + 2(\boldsymbol{m}, \nabla)\frac{\partial}{\partial t}$$
$$+ \frac{\partial^2}{\partial t^2}(\boldsymbol{n}, \nabla) - \frac{\partial^2}{\partial t^2} - 4(\boldsymbol{\Omega}, \boldsymbol{n})^2,$$

$$C = 4(\boldsymbol{\Omega}, \nabla)^2 + \Delta\frac{\partial^2}{\partial t^2},$$

$$D = -\frac{1}{\gamma - 1}\frac{\partial^4}{\partial t^4} - \frac{4\Omega^2}{\gamma - 1}\frac{\partial^2}{\partial t^2} - 4(\boldsymbol{\Omega}, \nabla)(\boldsymbol{\Omega}, \boldsymbol{n}) + C$$
$$- \frac{2}{r}\frac{\partial^2}{\partial t^2} - \frac{\partial^2}{\partial t^2}(\boldsymbol{n}, \nabla).$$

We can eliminate the function δ from (A1.1) if we multiply the first equation by the operator C and the second one by A and subtract the results:

$$[CA]\delta + (CB - AD)T = CF - AG. \tag{A1.2}$$

The approximate calculation of the commutator and repetition of the $[CA]$ procedure allows one to get the equation whose solution describes large-scale internal and sound waves

$$\left\{\left[-\frac{\gamma}{\gamma-1}\frac{\partial^2}{\partial t^2} - 1\right]\Delta + \frac{1}{\gamma-1}\frac{\partial^4}{\partial t^4} + \frac{2}{r}\frac{\partial^2}{\partial t^2} + \frac{\gamma}{\gamma-1}\frac{\partial^3}{\partial r \partial t^2}\right.$$
$$+ 2\frac{\partial}{\partial r}\frac{1}{r} + \frac{\partial^2}{\partial r^2} - \frac{4\gamma\Omega^2 \cos^2 v}{\gamma-1}\left(\frac{\partial^2}{\partial r^2} - \frac{\partial}{\partial r}\right) + \frac{2(2-\gamma)\Omega}{r(\gamma-1)}\frac{\partial^2}{\partial \varphi \partial t}$$
$$\left. + \frac{2\Omega}{r^2}\frac{\partial}{\partial \varphi}\int_0^t dt\right\}T = \int dt^5 \left[C^2 F + ACG - 2CAG\right] \equiv \mathcal{F}. \tag{A1.3}$$

The part of the operator in (A1.3) that contains Ω describes the Coriolis acceleration effects on an internal wave. The contributions proportional to $1/a$ allow one to take into account the effect of planet curvature. The equation describes short waves as well but the Coriolis effects and planet geometry contribute only at long (comparing with wavelength) distance.

Appendix 2

Nonlinear Terms for Interacting Modes of Poincaré and Rossby Waves at a Rotating Channel Surface

By definition (2.45)

$$A^n = -\frac{1}{H_0}\left(Y_n, \sum_{pq}(Y_p + rY_p')(Y_q + rY_q')\varphi_p\varphi_{qx}\right.$$
$$+ (Y_p + rY_p')Y_q'S\varphi_p\mu_{qx} + SY_p'(Y_q + rY_q')\mu_p\varphi_{qx}$$
$$+ S'^2\mu_p\mu_{qx} + Y_p\left(Y_q'(1+r_\beta) - (l^2 + \frac{\beta^2}{4})Y_q\right)\theta_p\varphi_q$$
$$\left. + SY_pY_q''\theta_p\mu_q\right), \quad Y_p' = \frac{\partial Y_p}{\partial y}. \quad (A2.1)$$

In correspondence with (2.44)

$$\mu_p = \sum_\alpha W_p^{(\alpha)}, \quad \varphi_p = \sum_\alpha a_p^{(\alpha)}\left(-i\frac{\partial}{\partial x}\right)W_p^{(\alpha)},$$
$$\theta_p = \sum_\alpha b_p^{(\alpha)}\left(i\frac{\partial}{\partial x}\right)W_p^{(\alpha)}. \quad (A2.2)$$

Here the mode indices $a_\alpha \to a_p^{(\alpha)}$, $b_\alpha \to b_p^{(\alpha)}$ are introduced. If we denote

$$(n,p,q) = (Y_n, Y_pY_q), \quad (n,p',q) = (Y_n, Y_p'Y_q),$$
$$(n,p',q') = (Y_n, Y_p'Y_q'), \quad (A2.3)$$

then by removing the brackets in (A2.1) and substituting (A2.2) we get

$$A^n = -\frac{1}{H_o}\sum_{pq\alpha\beta}\left\{[(n,p,q) + r[(n,p',q) + (n,p,q') + r(n,p',q')]]\right.$$
$$\times \left(a_p^{(\alpha)}W_p^{(\alpha)}\right)\left(a_q^{(\beta)}W_{qx}^{(\beta)}\right)$$
$$+ S[(n,p,q') + r(n,p',q')]W_p^{(\alpha)}W_{qx}^{(\beta)}$$
$$+ \left[(1+\beta r)(n,p,q') - (l_q^{(2)} + \frac{\beta^2}{4})(n,p,q)\right]\left(b_q^{(\alpha)}W_q^{(\alpha)}\right)\left(a_q^{(\beta)}W_{qx}^{(\beta)}\right)$$
$$\left. + S\left[\beta(n,p,q') - (l_q^2 + \frac{\beta^2}{4})(n,p,q)\right]\left(b_p^{(\alpha)}W_p^{(\alpha)}\right)W_q^{(\beta)}\right\}. \quad (A2.4)$$

We see that the nonlinearity is of mixed type: there are KdV terms as well as contributions of n-wave type. Analogously we have

$$B^n = -\frac{1}{H_o}\sum_{pq\alpha\beta}\left\{[(n,p,q') + r(n,p',q')]\left(a_p^{(\alpha)}W_p^{(\alpha)}\right)\left(b_q^{(\beta)}W_{qx}^{(\beta)}\right)\right.$$
$$+ S(n,p',q')W_p^{(\alpha)}b_q^{(\beta)}W_{qx}^{(\beta)}$$
$$\left.+ (n,p,q')\left(b_p^{(\alpha)}W_p^{(\alpha)}\right)\left(b_q^{(\beta)}W_q^{(\beta)}\right)\right\}, \tag{A2.5}$$

$$C_n = -\frac{1}{H_0}\sum_{pq\alpha\beta}\left\{\left[(n,p,q) + r\left[(n,p',q) + (n,p,q')\right] + r^2(n,p',q')\right]\right.$$
$$\times a_p^{(\alpha)}W_p^{(\alpha)}W_q^{(\beta)} + s\left[(n,p',q) + r(n,p',q')\right]W_p^{(\alpha)}W_q^{(\beta)}$$
$$- c_0^2 s\left[(n,p,q') + r(n,p',q')\right]\left(a_p^{(\alpha)}W_p^{(\alpha)}\right)\left(a_q^{(\beta)}W_q^{(\beta)}\right)$$
$$\left.- c_0^2 s^2(n,p',q')W_p^{(\alpha)}a_q^{(\beta)}W_q^{(\beta)}\right\}_x + C_L^n,$$

$$C_L^n = \beta\sum_{p,\alpha}\left\{(Y_n,yY_p')\,b_p^{(\alpha)}W_p^{(\alpha)} + (Y_n,y(Y_p+rY_p'))\,a_p^{(\alpha)}W_{px}^{(\alpha)}\right.$$
$$\left.+ S\left(Y_n,yY_p'\right)W_{px}^{(\alpha)}\right\}. \tag{A2.6}$$

Now we calculate the long waves limit in the projection operators (2.40,41). Omitting the mode index $pa_p^{(1)} = a_1\ldots$ we define the explicit form of $P^{(\alpha)}$ by the values of the functions a_i, b_i. Setting $k=0$ we get

$$a_1(0) = \frac{\beta}{f}\frac{c_0^2}{\beta^2 + l^2 + f^2/c_0^2}, \quad b_1(0) = 0, \tag{A2.7}$$

$$a_{2,3}(0) = \frac{f}{\beta}, \quad b_{2,3} = \pm c_0\sqrt{l^2 + f^0/c_0^2 + \beta^2/4}, \tag{A2.8}$$

$$\Delta_1 = 2b_2 f/\beta, \quad \Delta_2 = \Delta_3 = -b_2 a_1, \tag{A2.9}$$

$$\Xi = 2b_2(f/\beta - a_1). \tag{A2.10}$$

Thus, substituting the results into (2.40,41) we have

$$P_2 = \frac{1}{\Xi}\begin{pmatrix}\Delta_2 & b_1 - b_3 & a_3 - a_1 \\ a_2\Delta_2 & a_2(b_1-b_3) & a_2(a_3-a_1) \\ b_2\Delta_2 & b_2(b_1-b_3) & b_2(a_3-a_1)\end{pmatrix}$$
$$= \frac{1}{2(f/\beta - a_1)}\begin{pmatrix}-a_1 & 1 & (f/\beta - a_1)/b_2 \\ -a_1 f/\beta & f/\beta & f(f/\beta - a_1)/\beta b_2 \\ -b_2 a_1 & b_2 & f/\beta - a_1\end{pmatrix}$$

$$P_3 = \frac{1}{2(f/\beta - a_1)}\begin{pmatrix}-a_1 & 1 & (a_1 - f/\beta)/b_2 \\ -a_1 f/\beta & f/\beta & f(a_1 - f/\beta)/\beta b_2 \\ a_1 b_2 & -b_2 & a_1 - f/\beta\end{pmatrix}, \tag{A2.11}$$

$$P_1 = \frac{1}{f/\beta - a_1}\begin{pmatrix}f/\beta & -1 & 0 \\ a_1 f/\beta & -a_1 & 0 \\ 0 & 0 & 0\end{pmatrix}. \tag{A2.12}$$

Appendix 3

Projection to the Eigen Subspaces for Acoustic and Internal Waves

The linear equations for the two-dimensional wave disturbance of an exponential atmosphere lead to the dispersion relation (5.38) with the roots

$$\omega_{a,b}^2 = \frac{1}{2}\left[c^2\left(k^2 + 1/4H^2\right) \pm \sqrt{c^2(k^2 + 1/4H^2)^2 - 4\omega_g^2 c^2 k_x^2}\right]. \qquad (A3.1)$$

For every eigenvalue we find the eigen vector by the rules of Sect. 2.2, 3.5 which are described in detail for Poincaré and Rossby waves. As a result, we get the set of vectors for acoustic $\Phi_a^\pm$ and internal waves Φ_b^t. The signs $\pm$ denote right and left waves:

$$\Phi_a^\pm = \begin{pmatrix} 1 \\ \pm\alpha_a \\ \pm\beta_a \\ \gamma_a \end{pmatrix} \varphi_a^\pm, \quad \Phi_b^\pm = \begin{pmatrix} 1 \\ \pm\alpha_b \\ \pm\beta_b \\ \gamma_b \end{pmatrix} \varphi_b, \qquad (A3.2)$$

where

$$\alpha_a = \frac{\omega_a^2 + (ik_z - 1/2H)g}{H\omega_a(k^2 + ik_z/H - 1/4H^2)} k_x,$$

$$\beta_a = \frac{\omega_a^2(ik_z - 1/2H) + k_x^2 g}{iH\omega_a(k^2 - 1/4H^2 - ik_z/H)},$$

$$\gamma = \frac{\omega_a^2 - gH(k^2 + 1/4H^2)}{RH(k^2 - 1/4H^2 - ik_z/H)/\mu}.$$

The set of vectors $\Phi_{a,b}^\pm \equiv \Phi_i$ is complete. Solutions of wave equations in k-representation may be written in the form of the expansion $\Psi = \sum S_i \Phi_i$, where $S_i = C_i \exp(i\omega_i t)$. If the Fourier transform of an initial condition is

$$\Psi(k_x, k_z, 0) = \Psi_0(k),$$

then

$$C_1 = \sum_i \frac{F_i}{2(\omega_a^2 - \omega_b^2)} \Psi_{i0}(k) \sim \Phi_a^+,$$

where

Appendix 3

$$F_1 = -gH\left[k^2 - (1/4H^2) - \omega_b^2/gH\right],$$

$$F_2 = \frac{\omega_a}{k_x}\left[\frac{\omega_b^2}{g}(ik_z - 1/2H) - k_x^2\right],$$

$$F_3 = i\omega_a\left(\omega_b^2/g - ik_z - 1/2H\right),$$

$$F_4 = -gH\left[k^2 - (1/4H^2) + ik_z/H\right],$$

$$\Phi_a^- \sim C_2 = \left(F_1\Psi_1^0 - F_2\Psi_2^0 - F_3\Psi_3^0 + F_4\Psi_4^0\right)/2\left(\omega_a^2 - \omega_b^2\right),$$

$$\Phi_b^+ \sim C_3 = \sum_i \frac{G_i\Psi_{i0}(\mathbf{k})}{2(\omega_a^2 - \omega_b^2)},$$

where

$$G_1 = k^2 - \omega_a^2/(gH),$$

$$G_2 = \omega_b\left[\omega_a^2(ik_z - 1/2H)/(k_zg) - k_x\right],$$

$$G_3 = i\omega_b\left[(ik_z - 1/2H) - \omega_a^2/g\right],$$

$$G_4 = (k^2 - 1/4H^2 + ik_z/H)/T_0,$$

$$\Phi_b \sim C_4 = (G_1\Psi_{10} - G_2\Psi_{20} - G_3\Psi_{30} + G_4\Psi_{40})/2(\omega_a^2 - \omega_b^2).$$

The projection operators may be constructed by analogy with the Rossby and Poincaré case with the aid of the basis system (2.38) and are expressed by the coefficients in (A3.2) as in (A2.11), (2.39–41) following (3.68). The equations of the interaction are derived by the scheme presented in Sect. 2.3 [A.1].

Appendix 4

Basic Vectors, Interaction Operator for Atomic Nuclei with Spin 1

A Three-wave Resonance Condition for a Rectangular Waveguide. The energy operator for nuclei in the electric field of their election shells and in an external magnetic field directed along the z axis consists of two parts

$$\mathcal{H} = \mathcal{H}_Q + \gamma H_0 I_3 , \tag{A4.1}$$

where $\mathcal{H}_Q$ is given in Sect. 3.3 (3.38). To use the density matrix elements of (3.3) the basis for $\mathcal{H}$ and the interaction operator matrix have to be constructed in this basis. In this case it is necessary to diagonalize the operator $\mathcal{H}$ only in the subspace of the vectors $|\pm\rangle$, since the vector $|0\rangle$ is already an eigen vector. The substitution of (A4.1) into the spectral equation

$$\mathcal{H}|\xi\rangle = \mathcal{H}\left(\mu|+\rangle + \nu|-\rangle\right) = E\left(\mu|+\rangle + \nu|-\rangle\right) \tag{A4.2}$$

gives the expression for E in terms of the Hamiltonian parameters

$$E = A \pm \sqrt{\gamma^2 H_0^2 + 4B^2} , \tag{A4.3}$$

and also the relations

$$\nu_{1,2} = \frac{1}{\gamma H_0} \left(-2B \pm \sqrt{\gamma^2 H_0^2 + 4B^2}\right) \mu_{1,2} . \tag{A4.4}$$

The normalizing conditions $\langle 1|1\rangle = \langle 2|2\rangle = 1$ give

$$\mu_i = \frac{1}{\sqrt{1-\nu_i^2}} . \tag{A4.5}$$

The explicit form for $|\xi\rangle$ in $|\pm\rangle$ is obtained by means of the identities $\mu_1^2 + \mu_2^2 = 1$, $\nu_1 \nu_2 = -1$:

$$|1\rangle = \mu_1|+\rangle + \mu_2|-\rangle \quad , \quad |2\rangle = \mu_2|+\rangle - \mu_1|-\rangle . \tag{A4.6}$$

The form of the matrices I_k is derived by operating on (A4.6). If we denote $\varepsilon = \gamma H_0 / 2B$, then

Appendix 4

$$I_3 = \begin{pmatrix} 2\mu_1\mu_2 & \mu_2^2 - \mu_1^2 & 0 \\ \mu_2^2 - \mu_1^2 & -2\mu_1\mu_2 & 0 \\ 0 & 0 & 0 \end{pmatrix} ;$$

$$2\mu_1\mu_2 = \frac{\varepsilon}{\sqrt{1+\varepsilon^2}} ;$$

$$\mu_2^2 - \mu_1^2 = \frac{\sqrt{1+\varepsilon^2}}{\varepsilon^2/2 + 1} ;$$

$$I_2 = I_y = \begin{pmatrix} 0 & 0 & -i\mu_2 \\ 0 & 0 & i\mu_1 \\ i\mu_2 & -i\mu_1 & 0 \end{pmatrix} ; \quad I_1 \equiv I_x = \begin{pmatrix} 0 & 0 & \mu_1 \\ 0 & 0 & \mu_2 \\ \mu_1 & \mu_2 & 0 \end{pmatrix}. \quad (A4.7)$$

When the magnetic field $H_0 \to 0$, (A4.7) are transformed to (3.41) and $|1\rangle \to |+\rangle$, $|2\rangle \to |-\rangle$. The interaction operator matrix has the form

$$V = -\gamma \begin{pmatrix} 2H_z\mu_1\mu_2 & H_z(\mu_2^2 - \mu_1^2) & -i\mu_2 H_y + \mu_1 H_x \\ (\mu_2^2 - \mu_1^2)H_z & -2\mu_1\mu_2 H_z & i\mu_1 H_y + \mu_2 H_x \\ \mu_1 H_x + i\mu_2 H_y & \mu_2 H_x - i\mu_1 H_y & 0 \end{pmatrix}. \quad (A4.8)$$

By means of (3.39), interaction operators for higher spins can be written.

Let us consider the three-wave resonance and spatial synchronicity condition for the waveguide dispersion relation. The three-wave frequency resonance

$$\omega_1 = \omega_2 + \omega_3 \quad (A4.9)$$

may be obtained by the correct choice of frequencies without any difficulty. The simultaneous condition at synchronicity

$$k_1 = k_2 + k_3 \quad (A4.10)$$

requires a restriction to special mode numbers.

Suppose three dispersion relations are given

$$\omega_i^2 = \kappa_i^2 + k_i^2 \quad (A4.11)$$

and every one of them is specified by the pair m, n that determines κ_i by (3.17). We square (A4.9) and (A4.10). The result, together with (A4.11), gives

$$\kappa_1^2 - \kappa_2^2 - \kappa_3^2 + 2k_2k_3 = 2\sqrt{\kappa_2^2 + k_2^2}\sqrt{\kappa_3^2 + k_3^2} . \quad (A4.12)$$

Restricting ourselves to positive ω_i we square (A4.12) denoting $\kappa^2 = \kappa_1^2 - \kappa_2^2 - \kappa_3^2$. We get the quadratic equation for k_3:

$$4\kappa_2^2 k_3^2 - 4\kappa^2 k_2 k_3 - 4(\kappa_2^2 + k_2^2)\kappa_3^2 + \kappa^4 = 0 .$$

The discriminant of this equation $D = 4(k_2^2 + \kappa_2^2)(\kappa^4 - 4\kappa_2^2\kappa_3^2)$ is positive when $\kappa^4 > 4\kappa_2^2\kappa_3^2$, or squaring κ^2 we shall write the equality as the condition for $\kappa^2 = a$ ($\kappa_2^2 = b$, $\kappa_3^2 = c$):

$$a^2 - 2(b+c)a + b^2 + c^2 - 6bc > 0 .$$

The quadratic polynomial is positive outside the interval between the roots $a_{1,2} = b + c \pm 2\sqrt{2bc}$. Thus if $\kappa_2^2 + \kappa_3^2 - 2\sqrt{\kappa_2\kappa_3} > 0$, two intervals are possible

$0 < \kappa_1^2 < \kappa_2^2 + \kappa_3^2 - 2\sqrt{2\kappa_2\kappa_3}$,
$\kappa_1^2 > \kappa_2^2 + \kappa_3^2 + 2\sqrt{2\kappa_2\kappa_3}$.

Appendix 5

Nonlinear Internal Waves and Shear Flow Interaction Constants

Let $\Phi_z(z) = \Phi_0 e^{z/z_0}$ be the basis vector that enters into the derivative of the dynamical variable $u_0 = U\Phi(z)$ and therefore into (6.1). The normalization condition defines the constant

$$\Phi_0^2 = 1 \bigg/ \!\! \int_{-1/2}^{1/2} e^{2z/z_0} dz = \frac{\sinh^{-1} 1/z_0}{z_0} = 0.27 \quad \left(z_0 = \frac{1}{\pi}\right).$$

For the projection by (6.3–5) the scalar products are necessary:

$$(Z^\nu, e) = \frac{4\Phi_0 \nu^2}{\pi^2}\left[\frac{\cosh(1/z_0) - (-1)^\nu}{\nu^2 + (\pi^2 z_0^2)^{-1}}\right]; \tag{A5.1}$$

$$\sum_{\nu \neq 2}(Z^\nu, e) = \sum_{\nu \neq 2} \frac{4\pi^2}{\pi \sinh \pi} \frac{\cosh \pi - \nu^2(-1)^\nu}{(1+\nu^2)^2} \equiv S. \tag{A5.2}$$

The nonlinear constant n_U is proportional to the projection along e: $(Z^n)_{z_0}^2$. Then for the projection of the equation of the interaction of the mode n with the flow, the scalar product of $\varphi_n = 2Z_z^n Z^n$ and basis vectors will be needed:

$$(\varphi_n, e) = -8\Phi_0^2 (n\pi)^2 \frac{\sinh(1/2z_0)}{1/z_0^2 + (2n\pi)^2},$$

$$(\varphi^n, Z^\nu) = \frac{n\pi \delta_{\nu, 2n}}{\sqrt{2}},$$

$$\sum_{\nu \neq 2}(\varphi, Z^\nu)(Z^\nu, e) = \frac{k\pi}{\sqrt{2}}(Z^{2n}, e) = -8\Phi_0 \frac{n^2 \sinh(1/2z_0)}{(2n)^2 + (1/\pi^2 z_0^2)}. \tag{A5.3}$$

For $n = 1$ in (A5.3) we get 0. Then we have

$$\varphi_{n0} = -\frac{8\Phi_0 n^2 \pi^2 \sinh(1/2z_0)}{(1-S)\left[(2\pi n)^2 + (1/z_0^2)\right]} = -\frac{8\Phi_0 n^2 \sinh(\pi/2)}{(1-S)(2^2 n^2 + 1)},$$

$$\varphi_{n\nu} = \frac{n\pi \delta_{\nu, 2n}}{\sqrt{2}} - (e, Z^\nu)\varphi_{n0}. \tag{A5.4}$$

Formula (A5.4) is calculated using (A5.2). Thus the constant is equal to $n_U = 2\omega^2 (Z^1)_{z_0}^2 = 2\omega^2(\varphi^1, e) = -(32/5)\pi\omega^2/\cosh(\pi/2)$. For the computation of the constant n_τ which enters into (6.12) we shall find the products of the basis functions ΦZ^1 and $(\Phi Z^1)_z$ over the basis vector with $\nu = 1$. By the definition (6.5),

$$(Z^\mu \Phi)_\nu = \Phi_0 \left\{ \frac{e^{1/2z_0}(-1)^{\mu-\nu}/z_0 - e^{-1/2z_0}}{(\mu - \nu)^2 \pi^2 + 1/z_0^2} \right.$$
$$\left. - \frac{e^{1/2z_0}(-1)^{\mu+\nu}/z_0 - e^{-1/2z_0}}{\pi^2(\mu+\nu)^2 + 1/z_0^2} \right\},$$

$$(\Phi_z Z^\mu)_{z\nu} = \nu \pi^2 \Phi_0 \left\{ (\mu+\nu) \frac{e^{1/2z_0}(-1)^{\mu+\nu} - e^{-1/2z_0}}{(\mu+\nu)^2 \pi^2 + 1/z_0^2} \right.$$
$$\left. + \frac{e^{1/2z_0}(-1)^{\mu-\nu} - e^{-1/2z_0}}{(\mu-\nu)^2 \pi^2 + 1/z_0^2}(\mu-\nu) \right\}.$$

The mode μ generation intensity forced by the propagation of mode ν is given by the constant $(c_\nu = 1/\pi\nu)$

$$n_\nu^\mu = ik n_\nu c_\nu^2 (\Phi_z Z^\mu)_{z\nu} - ik n_\mu (Z^\mu \Phi)_\nu,$$

$$n_1^1 = ik\Phi_0 \left[\frac{(2/z_0)\sinh(1/2z_0)}{1/z_0^2 + \pi^2(\mu+\nu)^2} - \left(\frac{e^{1/2z_0}}{z_0} - e^{-1/2z_0} \right) \frac{4\pi^2 z_0^2}{4\pi^2 + 1/z_0^2} \right].$$

$$= \frac{4ik}{5\pi} \Phi_0 \left[\frac{1}{2} \sinh \frac{\pi}{2} - e^{\pi/2} + \frac{e^{-\pi/2}}{\pi} \right].$$

The second mode $\sim Z^2$ does not enter into the solution basis. The third mode generation constant is equal to

$$n_3^1 = ik\Phi_0 \left[\frac{4\sinh 1/2z_0}{3z_0} \cdot \frac{-8\pi^2 + 1/z_0^2}{(16\pi^2 + 1/z_0^2)(4\pi^2 + 1/z_0^2)} \right.$$
$$\left. - \left(\frac{e^{1/2z_0}}{z_0} - e^{-1/2z_0} \right) \frac{12}{(16\pi^2 + 1/z_0^2)(4\pi^2 + 1/z_0^2)} \right]$$

$$= -ik\Phi_0 \left[\frac{28}{255\pi} \sinh \frac{\pi}{2} - \frac{12}{85\pi^2} \left(\pi e^{\pi/2} - e^{-\pi/2} \right) \right];$$

$$\left| \frac{n_3^1}{n_1^1} \right| \approx 0.14.$$

References

Chapter 1

1.1 V.E.Zakharov, S.M.Manakov, S.P.Novikov J.P.Pitaevski: *Theory of Solitons. The Method of Inverse Problems* (Nauka, Moscow 1980); [English: Plenum, New York 1984]
1.2 J.Pedlosky: *Geophysical Fluid Dynamics* (New York, a.o. 1979)
1.3 G.B.Whitham: *Linear and Nonlinear Waves* (Wiley, New York 1974)
1.4 J.K.Engelbrecht, V.E.Fridman, E.N.Pelinovsky: "Nonlinear Evolution Equations" in *Pitman Res.Notes in Mathem. Series No.180* (Longman Scientific and Techn. Groups, London 1988)
1.5 G.G.Stokes: Camb.Trans. **8**, 441–473 (1847); Phil.Mag. 23 (3), 349-356; (1848)
1.6 J.Boussinesq: Comptes Rend. **72**, 755–759 (1871)
1.7 J.W.Miles: J.Fluid Mech. **106**, 131–147 (1981)
1.8 H.Segur, J.Hammack: J.Fluid Mech. **118**, 285–304 (1982)
1.9 R.Hide, R.J.Mason, R.A.Plumb: J.Atmos.Sci. **34**, 930–950 (1977)
1.10 A.C.Scott: *Active and Nonlinear Waves Propagation in Electronics* (Wiley-Interscience, New York 1970)
1.11 R.K.Bullough, P.J.Caudrey, J.C.Eilbeck, J.D.Gibbon: "A General Theory of Self-Induced Transparency", Optoelectronics, **6**, 121–140 (1974)
1.12 S.A.Akhmanov, B.A.Visloukh, A.S.Chirkin: Usp.Fiz.Nauk SSSR **149**, No.3, 449–509 (1986)
1.13 R.K.Bullough, P.J.Caudrey (eds.): *Solitons* (Springer, Berlin-Heidelberg 1980)
1.14 R.K.Dodd, J.C.Eilbeck, J.D.Gibbon, H.C.Morris: *Solitons and Nonlinear Wave Equations* (Academic, New York 1982)
1.15 J.Satsuma, S.Takeno, N.Wadati: "Recent Developments in Soliton Theory", Suppl.Progr. Theor.Phys. **94** (1988)
1.16 T.Taniuti, C.C.Wei: J.Phys.Soc Jpn. **24**, 941–946 (1968)
1.17 T.Taniuti: Suppl.Progr.Theor.Phys. **55**, 1–35 (1974)
1.18 K.Watanabe, T.Taniuti: J.Phys.Soc. Jpn. **42**, 1397–1403 (1977)
1.19 Y.Kodama, T.Taniuti: J.Phys.Soc. Jpn. **45**, 298–314; **47**, 1706–1716 (1978)
1.20 V.P.Maslov: *Resonance Processes in Wave Theory and Self-focusing*, Moscow Inst. of Electronic Eng. (1983)
1.21 V.P.Maslov: *Mathematical Aspects of Integral Optics*, Moscow Inst. of Electronic Eng. (1983)
1.22 S.Yu.Dobrokhotov, V.P.Maslov: *Soviet Science Review*, Vol. 3, (Overseas Publishing Association, Harwood 1982) pp.221–311
1.23 R.Grimshaw: Proc.Roy.Soc.Lond. **A368**, 359–375 (1979)
1.24 L.A.Ostrovsky: "Nonlinear Internal Ocean Waves" in *Nonlinear Waves* (Nauka, Moscow 1979) pp.292–323; Okeanologia USSR **18**, 181–191 (1978)
1.25 L.A.Ostrovsky, E.N.Pelinovsky: Prikl.Mekh.Mat. USSR **38**, 121–124 (1974)
1.26 Yu.Z.Miropolsky: *Internal Ocean Wave Dynamics* (Gidrometeoizdat, Leningrad 1981)
1.27 S.B.Leble: Izv.Akad.Nauk SSSR, Fiz.Atm.Okean **20**, 1199–1204 (1984)
1.28 A.A.Zaitsev, S.B.Leble: *Theory of Nonlinear Waves* (Kaliningrad University Press, Kaliningrad 1984)
1.29 A.Maxworthy, L.Redekopp, P.Weldman: Icarus **33** (1978)
1.30 P.Malanotte-Rizzoli: Adv.Geophys. **24**, 147–224 (1982)
1.31 R.Hirota, J.Satsuma: Phys.Lett. **A85**, 407–408 (1981)

1.32 S.P.Kshevetsky, S.B.Leble: Izv.Akad.Nauk SSSR, Fiz.Atm.Okean **21**, 170–176 (1985)
1.33 S.P.Kshevetsky, S.B.Leble: in *Waves and Diffraction, Proceedings of the IX USSR Symposium* Vol.2, (Tbilisi University Press, Tbilisi 1985), pp.57–59; Izv.Akad.Nauk SSSR, Mekh.Zhidk.Gaza **3**, 151–157 (1988)
1.34 S.P.Kshevetsky, N.A.Korneev, S.B.Leble, A.Ya.Shpilevoy: in *The Ionosphere and Solar-Terrestial Interactions* (Nauka, Alma-Ata 1985) pp.59–67
1.35 A.V.Tur, V.V.Yanovsky: Preprint Charkov Phys-Tech. Inst. Akad.Nauk UkSSR, No.83–36 (1983)
1.36 R.Dodd, A.Fordy: Phys.Lett. **A89**, 168 (1982)
1.37 A.V.Mikhailov, A.B.Shabat: Teor.Math.Phys.USSR, **62**, No.1, 47–65 (1985)
1.38 N.N.Bogolyubov, A.K.Prikarpatsky: Teor.Math.Phys.USSR, **67** No.3, 410–425 (1986)
1.39 V.P.Silin: *Parametric Interactions* (Nauka, Moscow 1973)
1.40 O.Philips: "Wave Interactions – Idea Evolution" in *Modern Hydrodynamics*, J.Fluid Mech. (special issue), **106** (1981)
1.41 A.A.Novikov: Izv. VUZov, Radiofiz. **19**, No.2, 321–328 (1976)
1.42 V.A.Borovikov, M.Ya.Kelbert: in *Waves and Diffraction, Proceedings of the IX USSR Symposium* Vol. I (Tbilisi University Press, Tbilisi 1985), pp.379–382
1.43 C.Frenzen, J.Kevorkian: Wave Motion **7**, 25–42 (1985)
1.44 M.Oikawa, N.Yajima: Suppl.Progr.Theor.Phys. **55**, 36–51 (1974)
1.45 A.I.Bobenko, V.B.Matveev, M.A.Salle: Dokl.Akad. Nauk SSSR **265**, 1357–1360 (1982)
1.46 V.D.Lipovsky: Izv. Akad. Nauk SSSR, Fiz.Atm.Okean **21**, 864–872 (1985)
1.47 V.L.Ginzburg, A.A.Rukhadze: *Waves in Magnetoactive Plasma* (Nauka, Moscow 1975)
1.48 O.M.Phillips: *The Dynamics of the Upper Ocean* (Cambridge University Press, 1966)
1.49 G.B.Whitham: Proc.Roy.Soc. **A299**, 6–25 (1967)
1.50 T.Benjamin: J.Fluid Mech. **25**, 241–270 (1966)
1.51 H.Ono: J.Phys.Soc. Jpn. **39**, 1082–1091 (1975)
1.52 S.V.Levikov: Okeanologia SSSR **16**, 968–974 (1976)
1.53 R.J.Joseph: J.Phys.A **10**, 1225–1227 (1977)
1.54 V.B.Matveev, M.A.Salle: Dokl. Akad.Nauk SSSR **261**, 533–538 (1981)
1.55 V.M.Babich, V.S.Buldyrev, I.A.Molotkov: *Space-Time Ray Method: Linear and Nonlinear Waves* (Leningrad University Press, Leningrad 1985)
1.56 Y.Kodama, A.Hasegawa: IEEE J. **QE–23**(5), 510–524 (1987)
1.57 V.V.Konotop, V.E.Vekslerchik: Phys.Lett.A **131**, 357–360 (1988)
1.58 J.Kubota, D.Ko, L.Dobbs: J.Hydronautics **12**, 157–165 (1978)
1.59 D.A.Vereschagin, S.B.Leble: Izv. Akad. Nauk SSSR, Fiz.Atm.Okean **6**, 815–820 (1987)
1.60 I.M.Moroz, J.Brindley: Proc.Roy.Soc.Lond. **A377**, 379–404 (1981)
1.61 V.I.Petviashvili: Physica **3D**, 329–334 (1981)
1.62 E.D.Belokolos, A.I.Bobenko, V.B.Matveev, V.Z.Enolsky: Usp.Mat.Nauk SSSR **41**, No.2(248), 3–42 (1986)
1.63 B.A.Dubrovin: Usp.Mat.Nauk SSSR **36**, No.2, 11–80 (1981)
1.64 H.Segur, A.Finkel: Stud.Appl.Math. **73**, 183–220 (1985)
1.65 S.B.Leble: Pure Appl.Geophys. **127**, No.2/3, 491–527 (1988)
1.66 L.A.Sakhnovich: *Nonlinear World. Proceedings of the IV International Workshop. Kiev, USSR 1989* Vol.2 (Naukova Dumka, Kiev 1989) 314–317
1.67 D.Yu.Manin, V.I.Petviashvili: J.Exp.Theor.Phys.Lett. **38**, No.9, 427–430 (1983)
1.68 A.D.Richmond, S.Matsushita: J.Geophys.Res. **80** (A9), 2839–2850 (1975)
1.69 V.P.Maslov, S.Yu.Dobrokhotov, G.A.Omelyanov: Usp.Mat.Nauk SSSR **36**, 63–126 (1981)
1.70 O.A.Tretyakov: Proc.Sino-British Joint Meeting on Optical Fiber Communication. Beijing, China 1986, p.333
1.71 O.A.Tretyakov: Radiotekhn.Elektr. **34**, 917–926 (1989)
1.72 V.P.Silin: *Introduction to the Kinetic Gas Theory* (Nauka, Moscow 1971)
1.73 V.E.Zakharov: J.Exp.Theor.Phys. **62**, 1745–1755 (1972)
1.74 V.P.Silin, V.T.Tikhonchuk: J.Exp.Theor.Phys. **81**, 2039–2051 (1981)
1.75 S.B.Leble, M.A.Salle: Dokl.Akad.Nauk SSSR **284**, 110–114 (1985)

1.76 V.B.Matveev: Lett.Math.Phys. **3**, 503–512 (1979)
1.77 M.Knudsen: *The Kinetic Theory of Gases. Some Modern Aspects* (London 1934)
1.78 E.Meyer, G.Sessler: Z.Phys. **149**, 15–39 (1957)
1.79 O.K.Buckner, J.H.Ferziger: Phys.Fluids, **9**, 2315–2322 (1966)
1.80 G.I.Grigoriev, O.N.Savina: Izv. VUZov, Radiofiz. **21**, 811–815 (1978)
1.81 D.A.Vereschagin, S.B.Leble: in *Predictions of the Ionosphere* (Nauka, Moscow 1982) pp.122–127
1.82 N.A.Fuchs: *Evaporation and Drop Growth in Gas Media* (Izd. AN SSSR, Moscow 1958)
1.83 R.Grimshaw: Stud.Appl.Math **56**, 241–266 (1977)
1.84 G.I.Barenblatt: Izv.Akad.Nauk SSSR, Fiz.Atm.Okean **13**, 845–849 (1977)
1.85 M.Knoll: "Feinstrukturen in der Jahreszeitlichen Sprungschicht im JASIN Gebiet", Ber.Inst. Meersek Christian-Albrecht Univ. Kiel, No.133, IV (1984)

Chapter 2

2.1 V.P.Silin: *Introduction to the Kinetic Gas Theory* (Nauka, Moscow 1971)
2.2 F.M.Kuni: *Statistical Physics and Thermodynamics* (Nauka, Moscow 1981)
2.3 G.B.Whitham: *Linear and Nonlinear Waves* (Wiley, New York 1974)
2.4 E.E.Grossard, W.H.Hooke: *Waves in the Atmosphere* (Elsevier, New York 1975)
2.5 A.S.Monin, A.M.Yaglom: *Statistical Hydromechnaics*, Vol.1 (Nauka, Moscow 1965)
2.6 U.Chamberlain: *Theory of Planetary Atmospheres* (Academic, New York 1078)
2.7 A.A.Zaitsev, S.B.Leble: *Theory of Nonlinear Waves* (Kaliningrad University Press, Kaliningrad 1984)
2.8 J.Pedlosky: *Geophysical Fluid Dynamics* (New York, a.o. 1979)
2.9 V.M.Babich, V.S.Buldyrev, I.A.Molotkov: *Space-Time Rau Method: Linear and Nonlinear Waves* (Leningrad University Press, Leningrad 1985)
2.10 D.A.Vereschagin, S.B.Leble: Izv.Akad. Nauk SSR, Fiz.Atm.Okean **6**, 815–820 (1987)
2.11 M.Janovich, P.Lyous: J.Fluid Mech., No.4, 773–786 (1975)
2.12 A.D.Richmond: J.Geophys.Res. **84**, 1880 (1979)
2.13 Z.L.Kobaladze, A.D.Pataraya, A.G.Hantadze: Izv.Akad. Nauk SSSR, Fiz.Atm.Okean **17**, 428–429 (1981)
2.14 F.S.Bessarab, S.P.Kshevetsky, S.B.Leble: in *Problems in Nonlinear Acoustics XI Proc. ISNA*, Novosibirsk, 1987, pp.144–148
2.15 Yu.Z.Miropolsky: *Internal Ocean Wave Dynamics* (Gidrometeoizdat, Leningrad 1981)
2.16 A.A.Zaitsev, S.B.Leble: *New Methods in Nonlinear Wave Theory* (Kaliningrad University Press, Kaliningrad 1987)
2.17 I.M.Moroz, J.Brindley: Proc.Roy.Soc.Lond. A**377**, 379–404 (1981)
2.18 S.B.Leble, M.V.Loktionov, A.Ya.Shpilevoy: in *Waves and Diffraction, Proc. of the IX USSR Symposium*, Vol.2 (Tbilisi University Press, Tbilisi 1985) pp.367–370
2.19 M.V.Vinogradova, O.V.Rudenko, A.P.Sukhorukov: *Wave Theory* (Nauka, Moscow 1979)
2.20 V.L.Ginsburg, A.A.Rukhadze: *Waves in Magnetoactive Plasma* (Nauka, Moscow 1975)
2.21 L.M.Brekhovskikh: *Waves in Stratified Media* (Nauka, Moscow 1973)
2.22 R.Hide, R.J.Mason, R.A.Plumb: J.Atmos.Sci. **34**, 930–950 (1977)
2.23 J.K.Engelbrecht, V.E.Fridman, E.N.Pelinovsky, "Nonlinear Evolution Equations in *Pitman Res.Notes in Mathem. Series No.180* (Longman Scientific and Tech. Groups, London 1988)
2.24 S.B.Leble: Izv.Akad. Nauk SSSR, Fiz.Atm.Okean **20**, 1199–1204 (1984)
2.25 S.P.Kshevetsky, S.B.Leble: Izv.Akad. Nauk SSSR, Fiz.Atm.Okean **21**, 170–176 (1985)
2.26 S.P.Kshevetsky, S.B.Leble: in *Waves and Diffraction, Proc. of the IX USSR Symposium* Vol. 2 (Tbilisi University Press, Tbilisi 1985) pp.57–59; Izv.Akad. Nauk SSR, Mekh.Zhidk. Gaza, **3**, 151–157 (1988)
2.27 A.Maxworthy, L.Redekopp, P.Weldman: Icarus 33 (1978)
2.28 P.Malanotte-Rizzoli: Adv.Geophys. **24**, 147–224 (1982)
2.29 R.R.Long: Tellus **17**, 46 (1965)

References

2.30 T.Benjamin: J.Fluid Mech. **25**, 241–270 (1966)
2.31 E.N.Pelinovsky, N.N.Romanova: Izv.Akad. Nauk SSR, Fiz.Atm.Okean **13**, 1169–1174 (1977)
2.32 Yu.D.Borisenko, A.G.Voronovich, A.I.Leonov, Yu.Z.Miropolsky: Izv.Akad. Nauk SSSR, Fiz. Atm.Okean **12**, 293–301 (1976)
2.33 A.I.Leonov, Yu.Z.Miropolsky: Izv.Akad. Nauk SSSR, Fiz.Atm.Okean **11** (1975)
2.34 L.A.Ostrovsky, E.N.Pelinovsky: Prikl.Mekh.Math. SSSR **38**, 121–124 (1974)
2.35 N.A.Korneev, S.P.Kshevetsky, S.B.Leble: in *Studies of the Dynamics of the Ionosphere* (Inst.Zemn.Magn.Ionosf.Raspr.Radiovoln Akad. Nauk SSSR, Moscow 1980) pp.109–123
2.36 N.A.Korneev, S.P.Kshevetsky, S.B.Leble: in *Studies of the Ionosphere* No.32 (Radio i Svyaz, Moscow 1982) pp.54–56
2.37 S.A.Ermakov, E.N.Pelinovsky: Izv.Akad. Nauk SSSR, Fiz.Atm.Okean **11**, 1055–1061 (1975)
2.38 I.V.Karpov, S.P.Kshevetsky: in *Predictions of the Ionosphere* (Nauka, Moscow 1982) pp.128–133
2.39 N.A.Korneev, S.B.Leble, A.Ya.Shpilevoy: in *Predictions of the Ionosphere* (Nauka, Moscow 1982) pp.117–122
2.40 A.D.Richmond, S.Matsushita: J.Geophys.Res. **80** (A9), 2839–3850 (1975)
2.41 V.M.Smertin, A.A.Namgaladze: Geomagn.Aeron. **21**, 302–308 (1981)
2.42 N.A.Korneev, S.B.Leble, A.Ya.Shpilevoy: in *Waves and Diffraction. Proc. VIII USSR Symposium* Vol.2, Lvov 1981 (Akad.Nauk SSSR, Moscow 1981) pp.86–89
2.43 S.P.Kshevetsky, S.B.Leble: Izv.Akad. Nauk SSSR, Fiz.Atm.Okean **21**, 1004–1007 (1985)
2.44 S.B.Leble, A.Ya.Shpilevoy: Geomagn.Aeron. No.2, 328 (1984)
2.45 S.B.Leble, L.A.Nazvalyan: in *Studies of the Ionosphere* No.43 (Radio i Svyaz, Moscow 1986)
2.46 S.P.Kshevetsky, N.A.Korneev, S.B.Leble, A.Ya.Shpilevoy: in *The Ionosphere and Solar-Terrestial Interactions* (Nauka, Alma-Ata 1985) pp.59–67
2.47 S.B.Leble, M.A.Loktionov, A.Ya.Shpilevoy: *Generation and Propagation of Internal Waves in a Spherical Layer* (Kaliningrad University Press, VINITI 1985) Report No. N 5280–85
2.48 S.P.Kshevetsky, S.B.Leble, L.A.Nazvalyan: in *Studies of the Ionosphere* (Radio i Svyaz, Moscow 1985)
2.49 A.D.Richmond: J.Geophys.Res. **83**, 4131–4145 (1978)
2.50 I.Tolstoy: Rev.Mod.Phys. **35**, 207–217 (1963)
2.51 S.H.Francis: J.Atm.Terr.Phys. **37**, 1011–1054 (1975)
2.52 N.M.Gavrilov: *The Heat Effect of Gravity Waves in the Upper Atmosphere* (Leningrad University Press, VINITI 7187–73)
2.53 R.L.Waltershield: Geophys.Res.Lett. **8**, 1235–1238 (1981)
2.54 A.M.Galper, V.G.Kirillov-Ugrjumov: JETP Lett. **30**, 631–633 (1979)
2.55 V.I.Krassovsky: Ann.Geophys. **33**, 347–356 (1977)
2.56 B.B.Kadomtsev, V.I.Petviashvili: Dokl.Akad. Nauk SSSR **192**, 753–756 (1970)
2.57 A.I.Leonov: Dokl.Akad. Nauk SSSR **229**, 820–824 (1976)
2.58 D.Yu.Manin, V.I.Petviashvili: J.Exp.Theor.Phys. Lett. **38**, No.9, 427–430 (1983)
2.59 R.Dodd, A.Fordy: Phys.Lett **A89**, 168 (1982)
2.60 B.A.Dubrovin: Ph.D. Thesis, Leningrad Department of the Mathematics Institute of the USSR Acad.Sci. (1984)
2.61 A.I.Bobenko: Preprint No.-10-86, Leningrad Department of the Mathematics Institute of the USSR Acad. Sci., (1986)
2.62 A.I.Bobenko, L.A.Bordag: J.Phys.A.Math.Gen. **22**, 1259–1274 (1989)
2.63 F.A.Grünbaum: Phys.Lett. **A139**, 146–155 (1989)
2.64 A.R.Osborne, L.Bergamasco: Physica D **18**, 26 (1986)
2.65 H.Segur, A.Finkel: Stud.Appl.Math. **73**, 183–220 (1985)
2.66 G.Reinish, J.C.Fernandez: Phys.Lett. **A67**, 259–262 (1978)
2.67 G.L.Lamb, Jr.: *Elements of Soliton Theory* (Wiley, New York 1980)
2.68 K.Dafermos: in *Nonlinear Waves*, ed. by S.Leibovich, A.R.Seebass (Cornell University Press, Ithaca, N.Y. 1974) pp.113–150
2.69 V.D.Lipovsky: Izv.Akad. Nauk SSSR, Fiz.Atm.Okean **21** 864–872 (1985)
2.70 V.D.Lipovsky: Dokl.Akad. Nauk SSSR **286**, 334–339 (1986)

2.71 V.S.Buslaev, L.D.Faddeev, L.A.Takhtajan: Physica D **18**, 255–256 (1986)
2.72 T.Taniuti: Suppl.Progr.Theor.Phys. **55**, 1–35 (1974)
2.73 V.E.Zakharov, S.V.Manakov, S.P.Novikov, J.P.Pitaevski: *Theory of Solitons. The Method of Inverse Problems* (Nauka, Moscow 1980); [English: Plenum, New York 1984]
2.74 V.I.Drobzhev: in *Wave Perturbation in the Atmosphere* (Nauka, Alma-Ata 1977) pp.37–43
2.75 S.B.Leble: *The Coupled KdV Integrability* (Kaliningrad University Press, VINITI Report No.2926.B87)
2.76 R.Hirota, J.Satsuma: Phys.Lett. **A85**, 407–408 (1981)
2.77 T.Yoneyama: Prog.Theor.Phys. **72**, 1081–1088 (1984)
2.78 P.K.Chandra, A.R.Choudry: J.Math.Phys. **29**, 843–850 (1988)
2.79 Z.L.Kobaladze, A.D.Pataraya, A.G.Hantadze: Izv.Akad. Nauk SSSR, Fiz.Atm.Okean **17**, 649–653 (1981)
2.80 V.I.Petviashvili: JETP Lett. **32**, 632–635 (1980)
2.81 J.C.Charney: J.Metal. **4**, 135–162 (1947)
2.82 M.Abramovitz, M.Stegun: *Handbook of Mathematical Functions* (National Bureau of Standards, Appl.Math. Series 1964)
2.83 L.D.Pogrebenko: Trud.Gidromet.Nauch.Issl.Inst. SSSR **242** (1982)
2.84 M.V.Babich: Funk.Anal.Pril. **19**, 53–55 (1985)
2.85 E.D.Belokolos, A.I.Bobenko, V.B.Matveev, V.Z.Enolsky: Usp.Mat. Nauk SSSR **41**, No.2(248), 3–42 (1986)
2.86 B.Jurco: Phys.Lett. **A138**, 497–501 (1989)
2.87 F.Verheest: J.Math.Phys. **29**, 109 (1988)

Chapter 3

3.1 Yu.V.Novozhilov, Yu.A.Yappa: *Electrodynamics* (Nauka, Moscow 1978)
3.2 M.Janovich, P.Lyous: J.Fluid Mech., No.4, 773–786 (1975); M.V.Vinogradova, O.V.Rudenko, A.P.Sukhorukov: *Wave Theory* (Nauka, Moscow 1979)
3.3 V.L.Ginsburg, A.A.Rukhadze: *Waves in Magnetoactive Plasma* (Nauka, Moscow 1975)
3.4 B.M.Mashkovtsev, K.N.Tsibirov, B.F.Emelin: *Waveguide Theory* (Nauka, Moscow 1966)
3.5 G.B.Whitham: *Linear and Nonlinear Waves* (Wiley, New York 1974)
3.6 V.P.Silin: *Parametric Interactions* (Nauka, Moscow 1973)
3.7 A.C.Scott: *Active and Nonlinear Waves Propagation in Electronics* (Wiley-Interscience, New York 1970)
3.8 S.V.Kuzkin, S.B.Leble: Fiz.Tverd.Tela SSSR **27**, 1483–1486 (1985)
3.9 U.Kh.Kopvillem, S.V.Prants: *Polarization Echo* (Nauka, Moscow 1975)
3.10 S.B.Leble: in *Coherent Excitation of Condensed Media* (Far East Scientific Center, Vladivostok 1979) pp.51–74
3.11 V.E.Zakharov, S.V.Manakov, S.P.Novikov, J.P.Pitaevski: *Theory of Solitons. The Method of Inverse Problems* (Nauka, Moscow 1980); [English: Plenum, New York 1984]
3.12 N.Blombergen: *Nonlinear Optics* (Wiley, New York 1965)
3.13 A.A.Zaitsev, S.B.Leble: *Theory of Nonlinear Waves* (Kaliningrad University Press, Kaliningrad 1984)
3.14 R.K.Dodd, J.C.Eilbeck, J.D.Gibbon, H.C.Morris: *Solitons and Nonlinear Wave Equations* (Academic, New York 1982)
3.15 J.Satsuma, S.Takeno, N.Wadati: "Recent Developments in Soliton Theory", Suppl. Progr. Theor.Phys. **94** (1988)
3.16 G.L.Lamb, Jr.: *Elements of Soliton Theory* (Wiley, New York 1980)
3.17 G.L.Lamb: Rev.Mod.Phys. **43**, 99–124 (1971)
3.18 D.J.Kaup: Phys.Rev.A **16**, 704–719 (1977)
3.19 D.J.Kaup, A.C.Newell: J.Math.Phys. **19**, 321–327 (1978)
3.20 M.G.Tseitlin: Teor.Mat.Phys. SSSR **57**, 238–248 (1983)
3.21 S.S.Iha: Pramana J.Phys. **11**, 313–322 (1978)

3.22 S.A.Akhmanov, B.A.Visloukh, A.S.Chirkin: Usp.Fiz.Nauk SSSR **149**, No.3, 449–509 (1986)
3.23 S.A.Akhmanov, R.V.Khohlov: *Nonlinear Optics Problems* (Itogi Nauki, Moscow 1964)
3.24 L.F.Mollenauer, R.H.Stolen, J.P.Gordon: Opt.Lett. **8**, 289–291 (1983); L.F.Mollenauer, R.H. Stolen: **9**, 13–14 (1984)
3.25 S.L.Palfrey, D.Grishkovsky: *Proc Conf. Lasers and Electro-Optics* (Baltimore, Maryland 1985) PDL.P.20
3.26 K.Ohkuma, , Y.H.Ichikawa, Y.Abe: Opt.Lett. **12**, 516–518 (1987)
3.27 A.Hasegawa, F.Tappert: Appl.Phys.Lett **23**, 142–144 (1973)
3.28 J.Satsuma, N.Yajima: Suppl.Progr.Theor.Phys. **55**, 284–306 (1974)
3.29 H.Eichorn: Inv.Probl. **1**(3), 193–198 (1985)
3.30 Y.Kodama, A.Hasegawa: IEEE J. **QE–23**(5), 510–524 (1987)
3.31 V.V.Konotop, V.E.Vekslerchik: Phys.Lett.A **131**, 357–360 (1988)
3.32 V.V.Konotop, V.E.Vekslerchik: Proc. URSI (Stockholm 1989)
3.33 V.A.Vysloukh, I.V.Cherednik: Dokl.Akad.Nauk SSSR **299**, 110–114 (1988)
3.34 A.Korpel, M.Chattarjee: Proc. IEEE **69**, 1539–1556 (1981)
3.35 V.I.Reshetsky: J.Phys.C **17**, 5887–5891 (1984)
3.36 R.J.Joseph: J.Phys.A **10**, 1225–1227 (1977)
3.37 V.B.Matveev, M.A.Salle: Dokl.Akad. Nauk SSSR **261**, 533–538 (1981)
3.38 E.D.Belokolos, A.I.Bobenko, V.B.Matveev, V.Z.Enolsky: Usp.Mat.Nauk SSSR **41**, No.2(248), 3–42 (1986)
3.39 H.Ono: J.Phys.Soc. Jpn. **39**, 1082–1091 (1975)
3.40 S.Yu.Dobrokhotov, V.P.Maslov: *Soviet Science Review*, Vol.3 (Overseas Publishing Association, Harwood 1982) pp.221–311
3.41 S.B.Leble: Izv.Akad.Nauk SSSR, Fiz.Atm.Okean **20**, 1199–1204 (1984)
3.42 V.D.Lipovsky: Izv.Akad.Nauk SSSR, Fiz.Atm.Okean **21**, 864–872 (1985)
3.43 R.Hirota, J.Satsuma: Phys.Lett.A **85**, 407–408 (1981)
3.44 N.N.Bogolyubov, A.K.Prikarpatsky: Teor.Math.Phys.USSR **67**, No.3, 410–425 (1986)
3.45 D.Mikalache, M.Bertolotti, C.Sibilia: Prog.Opt. **27**, 229–313 (1989)
3.46 D.Mihalache, R.G.Nazmitdinov, V.K.Fedyanin: Sov.J.Part.Nuclei (USA) **20**, 86 (1989)

Chapter 4

4.1 B.B.Kadomtsev: *Collective Phenomena in Plasma* (Nauka, Moscow 1976, 1988)
4.2 V.I.Petviahsvili: Fiz.Plasmy SSSR **1**, 28 (1975)
4.3 V.L.Ginsburg, A.A.Rukhadze: *Waves in Magnetoactive Plasma* (Nauka, Moscow 1975)
4.4 V.P.Silin: *Introduction to the Kinetic Gas Theory* (Nauka, Moscow 1971)
4.5 F.M.Kuni: *Statistical Physics and Thermodynamics* (Nauka, Moscow 1981)
4.6 Yu.V.Novozhilov, Yu.A.Yappa: *Electrodynamics* (Nauka, Moscow 1978)
4.7 A.A.Vlasov: J.Exp.Theor.Phys. **8**, 291 (1938)
4.8 L.D.Landau: J.Exp.Theor.Phys. **16**, 574 (1946)
4.9 V.P.Silin, V.T.Tikhonchuk: J.Exp.Theor.Phys. **81**, 2039–2051 (1981)
4.10 V.M.Babich, V.S.Buldyrev, I.A.Molotkov: *Space-Time Ray Method: Linear and Nonlinear Waves* (Leningrad University Press, Leningrad 1985)
4.11 I.V.Karpov, S.B.Leble, V.M.Smertin: Geomagnet.Aeron. SSSR, No.4, 672–673 (1983)
4.12 V.E.Zakharov, S.V.Manakov, S.P.Novikov, J.P.Pitaevski: *Theory of Solitons. The Method of Inverse Problems* (Nauka, Moscow 1980); [English: Plenum, New York 1984]
4.13 A.A.Zaitsev, S.B.Leble: *Theory of Nonlinear Waves* (Kaliningrad University Press, Kaliningrad 1984)
4.14 V.I.Petviashvili: Vopr.Teor.Plaz. **9**, No.11, 59–82 (1979)
4.15 L.M.Gorbunov, V.P.Silin: J.Exp.Theor.Phys. **47**, 203–210 (1964)
4.16 L.M.Gorbunov, A.M.Timerbulatov: J.Exp.Theor.Phys. **53**, 1494–1497 (1967)
4.17 V.P.Maslov: *Mathematical Aspects of Integral Optics* Moscow Institute of Electronic Engineering (1983)

4.18   S.Yu.Dobrokhotov, V.P.Maslov: *Soviet Science Review*, Vol.3 (Overseas Publishing Association, Harwood 1982) pp.221–311
4.19   V.E.Zakharov: J.Exp.Theor.Phys. **62**, 1745–1755 (1972)
4.20   E.S.Benilov : J.Exp.Theor.Phys. **88**, 120–128 (1985)
4.21   V.V.Gorev, A.S.Kingsep: J.Exp.Theor.Phys. **66**, 2048–2053 (1974)
4.22   N.A.Korneev, S.V.Kuzkin, S.B.Leble: *Proc. USSR Conf. on Nondestructive Probes* (Polytechnical Institute Press, Khabarovsk 1981) pp.66–68
4.23   V.I.Petviashvili, V.V.Yan'kov: Vopr.Teor.Plaz. SSSR, No.14, 3–55 (1986)
4.24   V.P.Silin *Parametric Interactions* (Nauka, Moscow 1973)
4.25   S.Putterman, J.Roberts: Phys.Rep. **168**, 209–263 (1988)
4.26   V.P.Bychenkov, V.P.Silin, S.A.Uryupin: Phys.Rep. **164**, 119–215 (1988)
4.27   V.B.Matveev: Lett.Math.Phys. **3**, 503–512 (1979)
4.28   A.I.Bobenko, V.B.Matveev, M.A.Salle: Dokl.Akad.Nauk SSSR **265**, 1357–1360 (1982)

Chapter 5

5.1   J.Pedlosky: *Geophysical Fluid Dynamics* (New York, a.o. 1979)
5.2   E.E.Gossard, W.H.Hooke: *Waves in the Atmosphere* (Elsevier, New York 1975)
5.3   A.D.Richmond: J.Geophys.Res. **84**, 1880 (1979)
5.4   Yu.Z.Miropolsky: *Internal Waves Dynamics in Ocean* (Gydrometeoizdat, Leningrad 1981)
5.5   V.M.Babich, V.S.Buldyrev, I.A.Molotkov: *Space-Time Ray Method: Linear and Nonlinear Waves* (Leningrad University Press, Leningrad 1985)
5.6   V.L.Ginsburg, A.A.Rukhadze: *Waves in Magnetoactive Plasma* (Nauka, Moscow 1975)
5.7   M.V.Vinogradova, O.V.Rudenko, A.P.Sukhorukov: *Wave Theory* (Nauka, Moscow 1979)
5.8   V.P.Maslov: *Mathematical Aspects of Integral Optics* (Moscow Institute of Electronic Engineering 1983)
5.9   S.B.Leble: Izv.Akad.Nauk SSSR, Fiz.Atm.Okean **20**, 1199–1204 (1984)
5.10   T.Benjamin: J.Fluid Mech. **25**, 241–270 (1966)
5.11   H.Ono: J.Phys.Soc. Jpn. **39**, 1082–1091 (1975)
5.12   S.V.Levikov: Okeanologiya SSSR **16**, 968–974 (1976)
5.13   R.J.Joseph: J.Phys.A **10**, 1225–1227 (1977)
5.14   Y.Kodama, M.Ablovitz, J.Satsuma: J.Math.Phys. **23**, 564–586 (1982)
5.15   A.I.Bobenko, V.B.Matveev, M.A.Salle: Dokl.Akad.Nauk SSSR **265**, 1357–1360 (1982)
5.16   A.A.Zaitsev, S.B.Leble: *New Methods in Nonlinear Wave Theory* (Kaliningrad University Press, Kaliningrad 1987)
5.17   A.A.Zaitsev, S.B.Leble: *Theory of Nonlinear Waves* (Kaliningrad University Press, Kaliningrad 1984)
5.18   J.K.Engelbrecht, V.E.Fridman, E.N.Pelinovsky, "Nonlinear Evolution Equations" in *Pitman Res.Notes in Math. Series* No.180 (Longman Scientific and Tech. Groups, London 1988)
5.19   H.Segur, J.Hammack: J.Fluid Mech. **118**, 285–304 (1982)
5.20   R.D.Lebbedev, A.O.Raduli: Preprint Inst.Theor.Expt.Phys. NM–16 (Moscow 1982)
5.21   G.B.Whitham: *Linear and Nonlinear Waves* (Wiley, New York 1974)
5.22   K.K.Tung, R.S.Ko, J.J.Chang: Stud.Appl.Math. **65**, 189–221 (1981)
5.23   V.D.Lipovsky: Izv.Akad.Nauk SSSR, Fiz.Atm.Okean **21**, 864–872 (1985)
5.24   V.B.Matveev, M.A.Salle: Dokl. Akad.Nauk SSSR **261**, 533–538 (1981)
5.25   M.Ablowitz, J.Kodama, J.Satsuma: "Direct and inverse scattering problems of the nonlinear intermediate long wave equation": Preprint Bell Labs. 1981.
5.26   R.I.Joseph, R.Egri: J.Phys.A **11**, L97–102 (1978)
5.27   E.D.Belokolos, A.I.Bobenko, V.B.Matveev, V.Z.Enolsky: Usp.Mat.Nauk SSSR **41**, No.2(248), 3–42 (1986)
5.28   G.A.Korn, T.M.Korn: *Mathematical Handbook* (McGraw-Hill, New York 1968)
5.29   R.K.Bullough, P.J.Caudrey (eds.): *Solitons* (Springer, Berlin-Heidelberg 1980)
5.30   N.N.Romanova: Izv.Akad.Nauk SSSR, Fiz.Atm.Okean **17**, 131–134 (1981)

5.31 V.P.Silin: *Introduction to the Kinetic Gas Theory* (Nauka, Moscow 1971)
5.32 F.M.Kuni: *Statistical Physics and Thermodynamics* (Nauka, Moscow 1981)
5.33 N.A.Fuks, A.G.Sutugin: *Highly Dispersive Aerosols* (Itogi Nauki, Moscow 1969)
5.34 I.N.Ivchenko: Izv.Akad.Nauk SSSR, Mekh.Zhidk.Gaza, No.3, 182–183 (1986)
5.35 A.Banankhan, S.K.Loyalka: Phys.Fluids 30, 56–64 (1987)
5.36 CIRA, *U.S. Standard Atmosphere* (US Government Print. Office, Washington, D.C. 1976)
5.37 D.A.Vereschagin, S.B.Leble: Izv.Akad.Nauk SSSR, Fiz.Atm.Okean 6, 815–820 (1987)
5.38 D.A.Vereschagin, S.B.Leble: in *Predictions of the Ionosphere* (Nauka, Moscow 1982) pp.122–127
5.39 N.A.Fuks: Evaporation and Drop Growth in Gas Media (Izd. AN SSSR, Moscow 1958)
5.40 P.L.Bhatnagar, E.P.Gross, M.Krook: Phys.Rev. 94, 511–525 (1954)
5.41 A.D.Richmond, S.Matsushita: J.Geophys.Res. 80 (A9), 2839–2850 (1975)
5.42 V.M.Smertin, A.A.Namgladze: Geomagn.Aeron. 21, 302–308 (1981)
5.43 S.B.Leble, M.A.Loktionov, A.Ya.Shpilevoy: *Generation and Propagation of Internal Waves in a Spherical Layer* (Kaliningrad University Press, VINITI, No.N5280–85)
5.44 A.A.Buzdin, S.B.Leble: Izv. VUZov SSSR, Ser.Fiz. 23, 126 (1980)
5.45 L.Sirovich, J.K.Thurber: J.Math.Phys. 10, 239 (1969)
5.46 A.K.Shchekin, S.B.Leble, D.A.Vereshchagin: *Introduction to Physical Kinetics of Rarefied Gases* (Kaliningrad University Press, 1990)

Chapter 6

6.1 J.Pedlosky: *Geophysical Fluid Dynamics* (New York, a.o. 1979)
6.2 U.Chamberlain: *Theory of Planetary Atmospheres* (Academic, New York 1978)
6.3 Yu.Z.Miropolsky: *Internal Waves Dynamics in Ocean* (Gydrometeoizdat, Leningrad 1981)
6.4 M.Knoll: "Feinstrukturen in der Jahreszeitlichen Sprungschicht im JASIN Gebiet", Ber.Inst. Meersek Christian–Albrecht Univ. Kiel, No.133, IV (1984)
6.5 N.M.Gavrilov: *The Heat Effect of Gravity Waves in the Upper Atmosphere* (Leningrad University Press, VINITI 7187–73)
6.6 R.L.Waltershield: Geoph.Res.Lett. 8, 1235–1238 (1981)
6.7 J.W.Miles: J.Fluid Mech. 16, 209–227 (1963)
6.8 H.Nakazava: Progr.Theor.Phys. 71, 906–916 (1984)
6.9 L.M.Brekhovskikh: *Waves in Stratified Media* (Nauka, Moscow 1973)
6.10 V.A.Borovikov: in *Waves and Diffraction, Proc. of the IX USSR Symposium*, Tbilisi, 1985, Vol.1 (Tbilisi University Press, Tbilisi 1985)
6.11 L.W.Howard: J.Fluid Mech. 16, 333–342 (1963)
6.12 L.M.Hocking, K.Stewartson: Proc.Royal Soc.London A 326, 289–299 (1972)
6.13 D.J.Benney, S.A.Marlowe: Stud.Appl.Math. 54, 181–205 (1975)
6.14 S.A.Marlowe: Quart.J.Roy.Meteor.Soc. 103, 769–783 (1977)
6.15 A.K.Liu, D.J.Benney: Stud.Appl.Math. 64, 247–249 (1981)
6.16 N.N.Romanova, V.Yu.Tseitlin: Izv.Akad.Nauk SSSR, Fiz.Atm.Okean 19, 796–806 (1983)
6.17 V.E.Zakharov: J.Exp.Theor.Phys. 62, 1745–1755 (1972)
6.18 M.Tomita: Suppl.Progr.Theor.Phys. N 79 (1984)
6.19 P.L.Kapitsa: Dokl.Akad.Nauk SSSR 15, 595–602 (1947)
6.20 F.Kh.Abdullaev: Suppl.Progr.Theor.Phys. 179, No.1 (1989)
6.21 V.P.Silin, V.T.Tikhonchuk: J.Exp.Theor.Phys. 81, 2039–2051 (1981)
6.22 A.S.Monin, A.M.Yaglom: *Statistical Hydrodynamics* Vol.1 (Nauka, Moscow 1965)
6.23 G.I.Barenblatt: Izv.Akad.Nauk SSSR, Fiz.Atm.Okean 13, 845–849 (1977)
6.24 A.V.Ivanov, L.A.Ostrovsky, I.A.Soustova, L.Sh.Tsimring: in *Lrge-Scale Wave Action on the Sea Surface* (Inst.Prikl.Fiz.Akad.Nauk SSSR, Gorky 1982) pp.75–85
6.25 A.A.Buzdin, S.B.Leble, V.V.Navrotsky: in *Hydrodynamic Field Investigations by Acoustic Methods* (Far East Scientific Center, Vladivostok 1983) pp.67–74

6.26 S.P.Kshevetsky, S.B.Leble, L.A.Nazvalyan: in *Studies of the Ionosphere* (Radio i Svyaz, Moscow 1985)
6.27 A.D.Richmond: J.Geophys.Res. **83**, 4131–4145 (1978)
6.28 D.A.Vereschagin, S.B.Leble: Izv.Akad.Nauk SSSR, Fiz.Atm.Okean **6**, 815–820 (1987)
6.29 G.I.Grigoriev, V.P.Dokuchaev: Geomagn.Aeron. **9**, 650–654 (1969)
6.30 B.N.Gershman: *Plasma Dynamics of the Ionosphere* (Nauka, Moscow 1974)
6.31 A.G.Hantadze: *Some Questions on Conductive Atmosphere Dynamics* (Metsniereba, Tbilisi 1973)
6.32 K.H.Joynez, E.C.Butcher: J.Atm.Terr.Phys. **42**, 455–460 (1980)
6.33 B.E.Brunelli, A.A.Namgaladze: *Ionosphere Physics* (Nauka, Moscow 1988)
6.34 V.M.Polyakov, V.V.Rybin: Geomagn.Aeron. **15**, 806–813 (1975)
6.35 I.V.Karpov, S.B.Leble, V.M.Smertin: Geomagn.Aeron. SSSR, No.4, 672–673 (1983)
6.36 I.V.Karpov, S.B.Leble: Izv. VUZov, Radiofiz. **26**, 1599–1601 (1983)
6.37 I.V.Karpov, S.B.Leble: Geomagn.Aeron. **26**, 234–237 (1986)
6.38 B.I.Tedd, H.J.Strangeways, T.B.Jones: J.Atmos.Terr.Phys. **46**, 109–117 (1984)
6.39 M.Abramovitz, M.Stegun: *Handbook of Mathematical Functions* (National Bureau of Standards, Appl.Math.Series, 1964)
6.40 S.B.Leble, M.A.Loktionov, A.Ya.Shpilevoy: *Generation and Propagation of Internal Waves in a Spherical Layer* (Kaliningrad University Press, VINITI Report No.N5280–85)
6.41 S.B.Leble, L.A.Nazvalyan: in *Studies of the Ionosphere* No.43 (Radio i Svyaz, Moscow 1985)
6.42 R.Grimshaw: Stud.Appl.Math **56**, 241–266 (1977)
6.43 V.I.Shrira: Dokl.Akad.Nauk SSSR **255**, 201–205 (1980)
6.44 R.Grimshaw: Stud.Appl.Math. **65**, 159–188 (1981)
6.45 R.Grimshaw: J.Fluid Mech. **115**, 347–377 (1982)
6.46 C.Garret, W.Munk: Ann.Rev.Fluid Mech. **11**, 339–369 (1978)
6.47 V.Malvestuto: Phys.Fluids **22**, 1862–1867 (1989)

Appendix 1

A.1 V.M.Babich, V.S.Buldyrev, I.A.Molotkov: *Space-Time Rau Method: Linear and Nonlinear Waves* (Leningrad University Press, Leningrad 1985)

Subject Index

Acoustic-gravitational waves   V, 1, 10
Acoustic (sound) waves   8, 18, 74, 82, 86, 105, 140
Alfven waves   81
Algebraic solutions   7
Algebraic solutions integration   7, 10
Algebro-geometrical integration   7
Ambipolar diffusion   11, 114, 122
Atmospheric quasiwaveguide propagation   106
Atmospheric wave   V1, 2, 15, 16, 17, 21, 45, 107, 136
Auroral electrojet   27, 33, 41

Background turbulence   14
Barotropic planetary wave   9, 45
Benjamin–Ono equation (BO)   2, 3, 10, 66, 92, 93, 95, 96, 101
Beta-plane approximation   20, 45
Boltzmann equation   13, 106, 107, 113, 123, 135
Boussinesq approximation   98
Boussinesq system   19
Brent-Väsäla frequency   36
Bubnov-Galerkin method   V

Cauchy problem for CKdV   37, 40
Cherenkov effect   73
Cherenkov electron heating   87
Closed waveguide   12
Collisionless magnetic hydrodynamics   74
Collisionless plasma   71
Collisionless regime   V, 1, 7, 8, 108
Coriolis effects   18, 140
Coriolis parameter   27, 45
Coupled Korteveg-de Vries equations (systems) (CKdV)   2–4, 8, 9, 12, 26, 39–45, 64, 75, 108
Coupled KdV equations with damping   34, 112
Coupled KP equations   36

Darboux–Matveev transform   10, 87–89
Debye radius   70
Density matrix   50, 51, 77, 145
Dielectric layer (slab)   7, 9, 65, 68

Dielectric waveguides   V, 1, 50, 62, 64
Diffusion effect   114
Diffusion rate   125
Diffusion velocity   13, 122
Dirac perturbation theory   125
Dirichlet problem for the Helmholz equation   67
Discrete Silin–Tikhonchuk equation   6, 10, 87, 89
Dispersion branch   4, 8, 11, 73, 96, 115
Dispersion coefficient   26
Dispersion constant   2, 43, 44
Dispersion relation(s)   4, 17, 19–22, 47, 55, 65, 73, 74, 76, 80, 81, 95, 99, 143, 146
Dissipative effects   18, 34, 132

Electromagnetic modes interaction   39
Electromagnetic wave (dispersion)   V, 1, 2, 9, 10, 19, 58, 61, 64, 114
Exponential atmosphere   17, 18, 20, 104, 143
Elementary modes   V, 1, 12

F-layer of the ionosphere   11
Fibers   V, 1, 7, 49, 64
Fine pycnoclyne structure   135, 137
Fine structure   11, 114, 115, 136
Finite-gap solution   7, 34, 49, 98
Free path length vector   108

Generalized KP equation   33
Geopotential   45
Global propagation problem   33
Gravitational field   V, 1

Height scale (homogeneous atmospheric height)   15
Helico-acoustic interaction   10, 79, 82
Helicoidal wave   82
Helmholtz equation   67, 105
Hirota–Satsuma equations   4, 45
Holography   64
Homogeneous collisionless plasma   71

Subject Index

Hydrodynamical distribution function 109
Hydrodynamics, magnetic 74
Hydrostatic approximation 8, 26, 27, 35, 36

Impurity transfer 122
Incompressibility 4, 5, 99, 101
Incompressible Boussinesq fluid 36
Incompressible fluid 94, 120, 129, 130
Incompressible fluid with dissipation 8, 86
Induced turbulence 14
Instability of a temperature field 11
Interaction of acoustic and internal waves 19–20
Intermode capture effect 26
Internal (gravity) wave V1, 1, 4, 6, 8, 10, 11, 14, 16–19, 27, 29, 48, 66, 94, 97, 105, 106, 124, 125, 127, 139
Inverse problem method 1, 36, 39, 40, 49
Ion-acoustic plasma waves interaction 39
Ion-acoustic wave 10, 69, 74–79, 81, 86
Ionospheric plasma 11, 81, 114, 122, 123
Ion wave 74–77
Irreversible relaxation 119
Iterations in equations 38, 39

Johnson equation 2, 37
Joseph equation 2, 10, 66, 93, 94, 103

Kadomtsev–Petviashvili equation (KP) 2, 7, 10, 32, 33, 98, 103
Kinetic description of plasma waves 69
Kinetic wave theory approach 108
Klein–Fock equation 66
Knudsen number (Kn) 10, 70, 79
Knudsen regime 10, 76, 106
Kolmogorov hypothesis 121
Korteweg-de Vries equation (KdV) 2, 10, 35, 38, 39, 98, 103, 142
KdV equation with damping 35, 112, 113

Landau damping 73, 92
Langmuir and ion-acoustic waves interaction 10, 84
Langmuir collapse 84
Langmuir frequency 72, 80
Langmuir turbulence (equations) 7, 69, 87
Langmuir wave 10, 69, 74–79, 85–91
Lax pair 26, 43
Liouville equation 63
Long and short wave interaction 119
Long internal wave 7, 8, 11, 26, 27, 29, 36
Long ion wave 78
Long surface wave 19

Long waves 2, 17, 25, 112, 115–117, 132, 142
Longitudinal coordinates 1, 7, 34
Lyapunov stability 79

Magnetic hydrodynamics 74
Magnetoactive plasma 81
Magnetoacoustic plasma wave 8
Magnetospheric (magnetic) substorm 28, 123
Maxwell equations 50, 52
Mean field 8, 14, 78, 115, 116, 118–120, 122, 129, 131, 133–135, 137
Mean flow 115
Mode coefficient functions 2
Mode lifetime 112
Mode subspace 21, 66
Model nonlinear (evolution) equation 1, 2, 26, 36, 39, 57, 67, 97
Modified incompressibility condition 11
Modified Zakharov equations 11
Multi-mode problem 3, 7, 26
Multi-mode wave 40
Multi-stream hydrodynamics 80
Multi-wave interacting mode 57
Multiple three-waves interaction equation 49

Near-integrable system V, 1, 41
Nonlinear constant (coupling coefficient) 2, 26, 29, 48, 49, 50, 77, 81, 85, 94, 103, 112, 118
Nonlinear dispersion relation 49
Nonlinear instability of explosive type 84
Nonlinear Schrödinger equation (NS) 57, 67, 115, 118
Nonlinear string equation 5
Nonlocal dispersion of internal waves 6, 10
Nonlocal Kadomtsev–Petviashvili equation 6, 94
Nonlocal Korteweg-de Vries equation (Joseph equation) 6, 64
Nonsingular perturbation theory V1, 11, 12, 14, 39, 45, 92, 130, 131
Nuclear quadrupole moment 58
Nuclear magnetic moment 58
NQR-NMR resonances 62

Ocean waves 21
Oceanic thermocline 98, 104, 122, 135
One-dimensional turbulence 85

Painlevé test 45
Parametric amplifying effect 63
Parametric weak plasma turbulence 10, 84
Perturbation theory 63
Plasma ambipolar diffusion (equation) 11

Subject Index

Plasma heating 114
Plasma turbulence 69, 87
Plasma wave V, 1, 2, 19, 76
Poincaré wave 8, 22, 24, 25, 64
Poisson bracket algebra of the dynamical variables 63
Polarization coupling 81
Primary turbulence 133
Projection operators 4–6, 8, 16, 19, 23, 37, 67, 95, 97, 99
Pseudodifferential operator 3, 66
Pseudospin 63
Pycnoclyne 6, 11, 36, 66

Quasi-single-mode perturbation 41
Quasineutral plasma 123
Quasisoliton 9, 12, 36, 40, 42
Quasiwaveguide 1, 3, 66
Quasiwaveguide propagation 93, 94, 97, 104

Ray approximation method 16
Real time holography 64
Recombination velocity 123
Rectangular waveguide with conducting walls 52
Relaxation process 59
Relaxation time approximation 108, 112
Resonance triads 47
Resonant nonlinear interaction 114, 133
Riemann theta-function 33, 34
Right(left)-travelling wave 5, 6, 54
Rossby wave 8, 9, 16, 17, 19, 22, 24–26, 39, 44, 64, 99, 114, 141, 143
Rossby wave on a sphere 45

Self-induced transparency equations 49, 82
Shear flow 11, 114, 115, 120, 127, 135
ShG (hyperbolic sine-Gordon equation) 9
Shock sound wave 36
Sine-Gordon (SG) equation 9, 49, 63
Single-directed wave 4
Single-mode long wave 109, 111
Silin–Tikhonchuk equation 10, 87, 89
Silin–Tikhonchuk soliton V, I
Sound (acoustic) wave 6, 17, 19, 82, 85, 92, 115, 138, 143
Spatial dispersion 70
Spatial synchronicity 146
Spectral densities 120
Spectral densities of acoustic waves 86
Spectral densities of electromagnetic field functions 82
Spectral turbulence 89
Spheroidal functions 46
Strongly coupled triad 9
Strongly inhomogeneous stratification 92
Sturm–Liouville problem for transverse variable 2, 45, 95
Surface wave 1, 96
Surface wave filtration 100
Symmetry of CKdV 41

Telegraph equation 112
Tensor of the field gradient 58
Temperature field instability 127
Thermocline 66, 100, 133, 137
Thermosphere V, 1
Thermospheric internal wave 27, 33, 123
Thermospheric waveguide 29, 35, 41, 123
Three-dimensional instability 33
Three-level system 61
Three-wave system 9, 48, 81
Three-wave interaction 10, 56, 58, 62
Three-wave resonance condition 9, 58, 145
Transverse coordinates 1, 2, 10, 34
Transverse electric wave 54
Transverse magnetic wave 54
Turbulent energy balance equation 120
Turbulent plasma V, 1, 69
Turbulent spectral densities 84
Turbulent spectrum 85, 86

Väsälä frequency 36, 99, 100, 103, 104, 118, 121, 129, 131
Vlasov equation 70, 80

Walquist–Estabrook analysis 44
Water waves 2, 4
Wave damping 130
Wave mode interaction(self-action)
Wave synchronicity 47
Weak turbulence in plasma 120
Weakly nonlinear long waves 24
Whistling atmospheric (spiral) waves 81
Whittaker functions 126, 127

Zakharov system 10, 11, 79, 119, 134, 135
Zonal flow 115

N. G. Chetaev
Theoretical Mechanics
Translated from the Russian by I. Aleksanova
1989. 407 pp. 190 figs. Hardcover DM 68,-
ISBN 3-540-51379-5

This university-level textbook reflects the extensive teaching experience of N. G. Chataev, one of the most influential teachers of theoretical mechanics in the Soviet Union. The mathematically rigorous presentation largely follows the traditional approach, supplemented by material not covered in most other books on the subject. To stimulate active learning numerous carefully selected exercises are provided. Attention is drawn to historical pitfalls and errors that have led to physical misconceptions.
Extensive appendices contain material from additional lectures on optics and mechnics analogies, Poincaré's equation and the special theory of elasticity.
Distribution rights for the socialist countries, India and Iran:
V/O "Mezhdunarodnaya Kniga", Moscow

D. Park, Williams College, Williamstown, MA
Classical Dynamics and Its Quantum Analogues
2nd enl. and updated ed. 1990. IX, 334 pp. 101 figs. Hardcover DM 78,- ISBN 3-540-51398-1

The primary purpose of this textbook is to introduce students to the principles of classical dynamics of particles, rigid bodies, and continuous systems while showing their relevance to subjects of contemporary interest. Two of these subjects are quantum mechanics and general relativity. The book shows in many examples the relations between quantum and classical mechanics and uses classical methods to derive most of the observational tests of general relativity. A third area of current interest is in nonlinear systems, and there are discussions of instability and of the geometrical methods used to study chaotic behaviour. In the belief that it is most important at this stage of a student's education to develop clear conceptual understanding, the mathematics is for the most part kept rather simple and traditional.
This book devotes some space to important transitions in dynamics: the development of analytical methods in the 18th century and the invention of quantum mechanics.

A. Hasegawa, AT & T Bell Laboratories, Murray Hill, NJ
Optical Solitons in Fibers
2nd enl. ed. 1990. XII, 79 pp. 25 figs.
Softcover DM 48,- ISBN 3-540-51747-2

Already after six months high demand made a new edition of this textbook necessary. The most recent developments associated with two topical and very important theoretical and practical subjects are combined: **Solitons** as analytical solutions of nonlinear partial differential equations and as lossless signals in dielectric **fibers.** The practical implications point towards technological advances allowing for an economic and undistorted propagation of signals revolutionizing telecommunications. Starting from an elementary level readily accessible to undergraduates, this pioneer in the field provides a clear and up-to-date exposition of the prominent aspects of the theoretical background and most recent experimental results in this new and rapidly evolving branch of science. This well-written book makes not just easy reading for the researcher but also for the interested physicist, mathematician, and engineer. It is well suited for undergraduate or graduate lecture courses.

A. G. Sitenko, Academy of the Ukrainian SSR

Scattering Theory

1990. Approx. 320 pp. 32 figs. (Springer Series in Nuclear and Particle Physics)
Hardcover DM 88,- ISBN 3-540-51953-X

This book is an introduction to nonrelativistic scattering theory. The presentation is mathematically rigorous, but is accessible to upper level undergraduates in physics. The relationship between the scattering matrix and physical observables, i.e. transition probabilities, is discussed in detail. Among the emphasized topics are the stationary formulation of the scattering problem, the inverse scattering problem, dispersion relations, three-particle bound states and their scattering, collisions of particles with spin and polarization phenomena. The analytical properties of the scattering matrix are discussed. Problems round off this volume.

B. N. Zakhariev, Moscow; A. A. Suzko, Minsk, USSR

Direct and Inverse Problems

Potentials in Quantum Scattering

1990. Approx. 200 pp. 42 figs.
Softcover DM 48,- ISBN 3-540-52484-3

This textbook can almost be viewed as a "how-to" manual for solving quantum inverse problems, that is, for deriving the potential from spectra or scattering data and also, as somewhat of a quantum "picture book" which should enhance the reader's quantum intuition. The formal exposition of inverse methods is paralleled by a discussion of the direct problem. Differential and finite-difference equations are presented side by side. The common features and (dis)advantages of a variety of solution methods are analyzed. To foster a better understanding, the physical meaning of the mathematical quantities are discussed explicitly. Wave confinement in continuum bound states, resonance and collective tunneling, energy shifts and the spectral and phase equivalence of various interactions are some of the physical problems covered.

P. C. Sabatier, University of Montpellier (Ed.)

Inverse Methods in Action

Proceedings of the Multicentennials Meeting, Montpellier, November 27th - December 1, 1989

1990. XIV, 636 pp. 125 figs.
Hardcover DM 138,- ISBN 3-540-51994-7

The basic idea of inverse methods is to extract from the evaluation of measured signals the details of the emitting them. The applications range from physics and engineering to geology and medicine (tomography).
Although most contributions are rather theoretical in nature, this volume is of practical value to experimentalists and engineers and as well of interest to mathematicians. The review lectures and contributed papers are grouped into eight chapters dedicated to tomography, distributed parameter inverse problems, spectral and scattering inverse problems (exact theory), wave propagation and scattering (approximations); miscellaneous inverse problems and applications and inverse methods in nonlinear mathematics.

Research Reports in Physics

The categories of camera-ready manuscripts (e.g., written in T_EX; preferably both hard and soft copy) considered for publication in the **Research Reports** include:

1. Reports of meetings of particular interest that are devoted to a single topic (provided that the camera-ready manuscript is received within four weeks of the meeting's close!).
2. Preliminary drafts of original papers and monographs.
3. Seminar notes on topics of current interest.
4. Reviews of new fields.

Should a manuscript appear better suited to another series, consent will be sought from the author for its transfer to the other series.

Research Reports in Physics are divided into numerous subseries, e.g., nonlinear dynamics or nuclear and particle physics. Besides covering material of general interest, the series provides the possibility for topics that are too specialized or controversial to be published within the traditional avenues. The small print runs make a consistent price structure impossible and will sometimes have to presuppose a financial contribution from the author (or a sponsor). In particular, in the case of proceedings the organizers are expected to place a bulk order and/or provide some funding.

Within **Research Reports** the timeliness of a manuscript is more important than its form, which may be unfinished or tentative. Thus in some instances, proofs may be merely outlined and results presented that will be published in full elsewhere later. Since the manuscripts are directly reproduced, the responsibility for form and content is mainly the author's.

☐ Heidelberger Platz 3, W-1000 Berlin 33, F. R. Germany ☐ 175 Fifth Ave., New York, NY 10010, USA ☐ 8 Alexandra Rd., London SW19 7JZ, England ☐ 26, rue des Carmes, F-75005 Paris, France ☐ 37-3, Hongo 3-chome, Bunkyo-ku, Tokyo 113, Japan ☐ Room 701, Mirror Tower, 61 Mody Road, Tsimshatsui, Kowloon, Hong Kong ☐ Avinguda Diagonal, 468-4° C, E-08006 Barcelona, Spain

Research Reports in Physics

Manuscripts should be no less than 100 and no more than 400 pages in length. They are reproduced by a photographic process and must therefore be typed with extreme care. Corrections to the typescript should be made by pasting in the new text or painting out errors with white correction fluid. The typescript is reduced slightly in size during reproduction; the text on every page has to be kept within a frame of 16 × 25.4 cm (6 5/16 × 10 inches). On request, the publisher will supply special stationery with the typing area outlined.

Editors or authors (of complete volumes) receive 5 complimentary copies and are free to use parts of the material in later publications.

All manuscripts, including proceedings, must contain a subject index. In the case of multi-author books and proceedings an index of contributors is also required. Proceedings should also contain a list of participants, with complete addresses.

Our leaflet, *Instructions for the Preparation of Camera-Ready Manuscripts,* and further details are available on request.

Manuscripts (in English) or inquiries should be directed to

Dr. Ernst F. Hefter
Physics Editorial 4
Springer-Verlag, Tiergartenstrasse 17
W-6900 Heidelberg, Fed. Rep. of Germany
(Tel. [0] 6221-487495;
Telex 461 723;
Telefax [0] 6221-413982)

☐ Heidelberger Platz 3, W-1000 Berlin 33, F. R. Germany ☐ 175 Fifth Ave., New York, NY 10010, USA ☐ 8 Alexandra Rd., London SW19 7JZ, England ☐ 26, rue des Carmes, F-75005 Paris, France ☐ 37-3, Hongo 3-chome, Bunkyo-ku, Tokyo 113, Japan ☐ Room 701, Mirror Tower, 61 Mody Road, Tsimshatsui, Kowloon, Hong Kong ☐ Avinguda Diagonal, 468-4° C, E-08006 Barcelona, Spain

Printing: Mercedesdruck, Berlin
Binding: Buchbinderei Lüderitz & Bauer, Berlin